Bill Ryder

NEXUS

NEXUS

*Small Worlds
and the Groundbreaking
Science of Networks*

MARK BUCHANAN

W. W. NORTON & COMPANY

New York   London

For information about permission to reproduce selections from this book, write to Permissions,
W. W. Norton & Company, Inc., 500 Fifth Avenue, New York, NY 10110

The text of this book is composed in Minion
with the display set in Aperto
Composition by Carole Desnoes
Manufacturing by The Haddon Craftsmen, Inc.
Book design by JAM Design
Production manager: Andrew Marasia
Page makeup by Carole Desnoes

Library of Congress Cataloging-in-Publication Data

Buchanan, Mark.
Nexus : small worlds and the groundbreaking science of networks
/ Mark Buchanan.— 1st ed.
p. cm.
Includes bibliographical references and index.
ISBN 0-393-04153-0 (hardcover)
1. Causality (Physics) 2. Pattern formation (Physical sciences)
3. Social networks. I. Title.
QC6.4.C3 B827 2002
530'.01—dc21
2002000518

W. W. Norton & Company, Inc., 500 Fifth Avenue, New York, N.Y. 10110
www.wwnorton.com

W. W. Norton & Company Ltd., Castle House, 75/76 Wells Street, London W1T 3QT

1 2 3 4 5 6 7 8 9 0

To Kate

CONTENTS

Acknowledgments   9

Prelude   11

1. Strange Connections   23

2. The Strength of Weak Ties   34

3. Small Worlds   48

4. Brain Works   61

5. The Small-World Web   73

6. An Accidental Science   89

7. The Rich Get Richer   106

8. Costs and Consequences   121

9. The Tangled Web   138

10. Tipping Points   156

11. Breaking Out, Small-World Style   170

12. Laws for the Living   184

13. Beyond Coincidence   197

Notes   209

Index   223

ACKNOWLEDGMENTS

I am grateful to the many scientists who have generously supplied me with figures, data, or useful bits of information crucial to the writing of this book. My thanks go out especially to Duncan Watts, Steve Strogatz, Thomas Blass, Mark Granovetter, Albert-László Barabási, Hawoong Jeong, Jack Scannell, Harry Swinney, Nebojša Nakićenović, Luís Amaral, Gene Stanley, Peter Yodzis, Gennady Gorelik, John Potterat, Alden Klovdahl, William Darrow, Alessandro Vespignani, Paul Meakin, Jean-Philippe Bouchaud, Marc Mézard, Hal Cheswick, Daniel Goldman, David Lavigne, Valeri Makarov, Manuel Velarde, Bill Krantz, Nigel Gilbert, Klaus Troitzsch, Brad Werner, and Mark Kessler. No one should suppose that any of these researchers agrees with everything or even most of what I have written here. Any errors or misconceptions that may have slipped into the text are mine alone.

I would like also to thank my agent Kerry Nugent Wells and my editor Angela von der Lippe at W. W. Norton for their faith in the value of this project and their great help in improving the book. Most importantly, infinite thanks go to my wife Kate for her unflagging support and encouragement during the many months of writing.

MARK BUCHANAN
Notre Dame de Courson
November 2001

PRELUDE

Science is built up with facts, as a house is with stones. But a collection of facts is no more a science than a heap of stones is a house.

—*Henri Poincaré*[1]

FORTY-ONE YEARS AGO, at the height of the Cold War, the philosopher Karl Popper published a short anti-Marxist volume entitled *The Poverty of Historicism*. In using the term *historicism*, Popper meant to refer to any system of ideas that claimed, like the philosophy of Karl Marx, that the unfolding of human history can be predicted in advance. Marx had famously asserted that communism was the world's social and political destiny. And Popper, possessor of a lifelong revulsion toward communism, aimed to take the wind out of Marx's claim.

Popper's argument was as clever as it was simple. To begin with, he said, we all accept that the growth of human knowledge has an influence on the course of history. In the 1930s scientists came to understand the basic physics of the atomic nucleus, and mankind soon had to face up to the disconcerting power of nuclear weapons. Changes in knowledge clearly have an effect on history. It is also true, said Popper, that we cannot predict how our knowledge will grow, for learning means discovering something new and unexpected. If we could predict future discoveries now, we would know about them already.

So, if changes in knowledge influence the course of history, and we cannot foresee such changes, history must be beyond prediction. "The belief in historical destiny," as Popper put it, "is sheer superstition. . . . There can be no prediction of the course of human history by scientific or any other rational methods."[2]

Whether this argument makes sense or not, most of us would accept its conclusion. Humanity is an immensely complicated network of more than six billion individuals, and in view of the formidable complexity of even a single human being, it is no surprise that our collective future cannot be foreseen. There are certainly no equations for history. Indeed, while the physical sciences reveal numerous regularities that can be captured in immutable scientific laws, this does not seem to be the case in the social world where emotional and unpredictable humans take center stage. Lump together all the fields that deal with the lives and actions of people—from history and economics to political science and psychology—and it is impossible to find a single topic that can be wrapped up with a few simple laws like those of physics or chemistry.

Is it conceivable that there could be mathematical laws for the human world? Many people find it distinctly unsettling even to consider the possibility. As individuals, we prize our freedom to do and think what we will. Mathematics, by contrast, is thought of as being rigid and restrictive, its inflexible symbols suited perhaps to the description of mindless and soulless matter, but certainly not to the lives of living, breathing humans.

One of the messages of this book, however, is that it may be possible to discover mathematical laws and meaningful patterns in the human world. The purpose of science, as the late social and political scientist Herbert Simon once put it, "is to find meaningful simplicity in the midst of disorderly complexity."[3] And over the past five years, sociologists, physicists, biologists, and other scientists have turned up numerous unexpected connections between the workings of the human world and the functioning of other seemingly unrelated things: from the living cell and the global ecosystem to the Internet and the human brain. This is not to say that we lack free will, or that Karl Popper was wrong and history can be predicted. But it does suggest that many of the inherent complexities of human society actually have little to do with the complex psychology of humans; indeed, similar patterns turn up in many other settings where conscious beings play no role at all.

Surprisingly, these discoveries were kicked off originally by studies in pure mathematics. Nevertheless, they are now offering insights into long-standing problems in diverse fields of science and into some of the oldest puzzles of human society.

SMALL WORLDS

ONE DAY DURING the winter of 1998, mathematicians Duncan Watts and Steve Strogatz of Cornell University in Ithaca, New York, sat down at a table in Strogatz's office and drew a series of dots on a piece of paper. They then connected some of the dots together with lines to produce a simple pattern that mathematicians refer to as a *graph*. This may not sound like serious mathematics; it certainly does not sound like a profitable way to go about making discoveries. But as the two mathematicians were soon to learn, they had connected their dots in a peculiar way that no mathematician had ever envisioned. In so doing, they had stumbled over a graph of an unprecedented and fascinating kind.

Watts and Strogatz came upon their graph while trying to make sense of a curious puzzle of our social world. In the 1960s, an American psychologist named Stanley Milgram tried to form a picture of the web of interpersonal connections that link people into a community. To do so, he sent letters to a random selection of people living in Nebraska and Kansas, asking each of them to forward the letter to a stockbroker friend of his living in Boston, but he did not give them the address. To forward the letter, he asked them to send it only to someone they knew personally and whom they thought might be socially "closer" to the stockbroker. Most of the letters eventually made it to his friend in Boston. Far more startling, however, was how quickly they did so—not in hundreds of mailings but typically in just six or so. The result seems incredible, as there are hundreds of millions of people in the United States, and both Nebraska and Kansas would seem a rather long way away -in the social universe—from Boston. Milgram's findings became famous and passed into popular folklore in the phrase "six degrees of separation." As the writer John Guare expressed the idea in a recent play of the same name: "Everybody on this planet is separated by only six other people. . . . The president of the United States. A gondolier in Venice. . . . It's not just the big names. It's anyone. A native in the rain forest. A Tierra del Fuegan. An Eskimo. I am bound to everyone on this planet by a trail of six people. It's a profound thought. . . ."[4]

It is a profound thought, and yet it really seems to be true. A few years ago a German newspaper accepted the light-hearted challenge of trying to connect a Turkish kebab-shop owner in Frankfurt to his favorite actor, Marlon Brando. After several months, the staff of *Die Ziet* discovered that it took no more than six links of personal acquaintance to do so. The kebab-shop owner, an Iraqi immigrant named Salah Ben Ghaln, has a friend living in California. As it happens, this friend works alongside the boyfriend of a woman who is the sorority sister of the daughter of the producer of the film *Don Juan de Marco*, in which Brando starred. Six-degrees of separation is an undeniably stunning characteristic of our social world, and numerous more careful sociological studies offer convincing evidence that it is true—not only in special cases, but generally. But how can it be true? How can six billion people be so closely linked?

These are the questions that Watts and Strogatz set for themselves. If you think of people as dots, and links of acquaintance as lines connecting them, then the social world becomes a graph. So for months Watts and Strogatz had been drawing graphs of all sorts, connecting dots in different patterns, hoping to find some remarkable scheme that would reveal how six billion people could conceivably be connected together so closely. They tried drawing dots and linking them together into a grid, arranged regularly like the squares on a chessboard. They tried drawing dots and connecting them haphazardly to produce random graphs, each looking like a connect-the-dots game gone haywire. But neither the ordered nor the random graphs seemed to capture the nuances of real social networks. The small-world mystery remained defiant.

Then finally, on that winter's day in 1998, the two researchers stumbled over their peculiar graph. What they discovered was a subtle way of connecting the dots that was neither orderly nor random but somewhere in-between, an unusual pattern with chaos mingling in equal balance with order. Playing with variations of this odd-looking graph over the next few weeks, Watts and Strogatz found that it held the key to revealing how it is that six billion people can be connected by only six links.

In this book, we will explore these amazing "small-world" graphs in more detail and see precisely how they work their magic. But these intriguing mathematical structures were merely the prelude to a far more important discovery. Curious to see how social networks are distinguished from other kinds of networks, Watts and Strogatz set to work studying the network of power lines in the United States and the network of neurons in the nematode worm, a creature so simple that biologists in the 1980s were able to work out a map of its entire nervous system. The U.S. power grid has been fashioned by human design; the nervous system of the worm, by evolution. Nevertheless, these networks turned out to have almost exactly the same small-world structure as the social world. For some mysterious reason, Watts and Strogatz's odd-looking graphs seemed to be pointing toward some deep organizing principle of our world.

In the few years since Watts and Strogatz published their initial discoveries, an explosion of further work by other mathematicians, physicists, and computer scientists has turned up profoundly similar structures in many of the world's other networks. Social networks turn out to be nearly identical in their architecture to the World Wide Web, the network of Web pages connected by hypertext links. Each of these networks shares deep structural properties with the food webs of any ecosystem and with the network of business links underlying any nation's economic activity. Incredibly, all these networks possess precisely the same organization as the network of connected neurons in the human brain and the network of interacting molecules that underlies the living cell.

These discoveries are making a new science of networks possible, a science that is the focus of this book. Surprisingly, both in the physical world and in the world of humans, the very same principles of design seem to be at work. Networks that have grown up under different conditions to meet markedly different needs turn out to be almost identical in their architecture. Why? A new theoretical perspective on networks is helping to answer this question and is enabling researchers in almost every area of science to begin tackling some of their most challenging and important problems.

For centuries scientists have been taking nature apart and analyzing its pieces in ever-increasing detail. By now it is hardly necessary to point out that this process of "reduction" can take understanding only so far. Learn all you want about the structure and properties of a single water molecule, for example, and you will still have no inkling that a collection of them will be a liquid at 1°C and a solid at 1°C. This abrupt change in state involves no alteration of the molecules themselves, but rather a transformation in the subtle organization of the network of their interactions. In an ecosystem or economy, the same distinction holds true. No amount of information at the level of the individual species or economic agent can hope to reveal the patterns of organization that make the collective function as it does. Today, the most fascinating and pressing problems almost invariably center on efforts to unravel the delicate and intricate organization of networks of bewildering complexity.

LINKS AND CONNECTIONS

IN FEBRUARY 2001, an international consortium of biologists revealed the completion of a "working draft" of the human genome, a more or less complete map of the genetic information contained in human DNA. This momentous achievement will trigger a terrific advance in the understanding of human diseases, and yet the genome is really only one step toward understanding what makes us human. Surprisingly, the Human Genome Project discovered that each of us has around thirty thousand genes, far below the roughly one hundred thousand that had been expected. This is particularly puzzling since some plants have nearly twenty-five thousand genes. Either we humans are not quite as sophisticated as we like to think, or it is not *merely* the number of genes that determines the complexity of an organism.

No liver or heart or brain is built from genes; rather, each gene contains instructions for making molecules known as *proteins*, which then take their place in a web of tens of thousands of other different proteins, all interacting with one another in complicated ways.[5] To comprehend what makes us alive, and especially what distinguishes us from plants, will require insight into the architecture of this vast network; our sophistication is not due to one or another protein, but to the delicate design of the entire network.

In the world of ecology, researchers face a similar struggle to deal with networks of mind-numbing complexity. As an example, the fishing industry in South Africa has long argued that culling seals off the west coast would increase the number of hake, a popular commercial catch. Seals eat hake, and so the fisheries' argument seems to possess a brute mathematical logic. But things are not that simple. Seals and hake are just two members of an immensely complicated food web (Figure 1), and no act of ecological meddling can be isolated. Ecologist Peter Yodzis of the University of Guelph in Canada estimates that a change in the number of seals would influence the hake population by acting through intermediate species in more than 225 *million* domino-like pathways of cause and effect.[6] Would culling seals make for more hake? As of now, no one can even hazard an intelligent guess. If the fisheries slaughtered seals, there might be fewer hake than before.

The seal-hake network is merely an example of the forbidding complexity of our ecosystem; there are others, of course, where a lack of understanding could have truly disastrous consequences. Through geo-

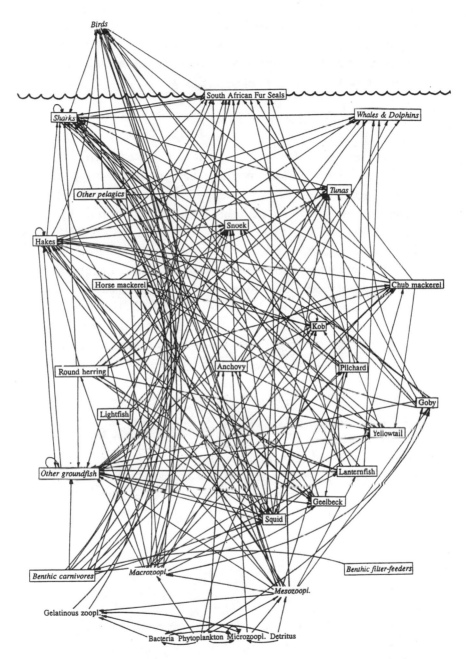

Figure 1. A portion of the food web for the Benguela ecosystem, which is located off the western coast of South Africa. (Reprinted by permission of Peter Yodzis.)

logical history there have been at least five great episodes of mass extinction, in each of which more than 50 percent of all species worldwide were suddenly eradicated. In recent years some scientists have suggested that we may now be in the midst of a sixth great extinction, this one being triggered by our own perturbation of the earth's environment. To judge the likelihood of such a tragic scenario, and to learn more about how we might avoid it, scientists need a better handle on the workings of complex networks.

When it comes to the social world, our understanding of complex webs of cause and effect is again sadly lacking. Take economics as an example. In every nation on the planet, the distribution of wealth among people is decidedly skewed, with a small fraction possessing almost all of it. This basic truth has been known for more than a century. What causes it? Is there something deeply rooted in capitalist economic principles that leads to such wealth concentration? Or is there something in human nature? Does the distribution of wealth reflect the distribution of wealth-accumulating talents in people? Although economists of differing political persuasions argue vociferously and emotionally for one point of view or another, orthodox economic theory has little to say about the matter. Lacking any way even to begin to understand an economy as a complex, evolving network, economic theory seems unable to explain one of the most universal and socially important facts of economic reality.

Clearly, what economists, ecologists, and biologists need is a way of gaining insights into the structures and workings of complicated networks. They need a theory that would offer some understanding regardless of the specific components that make up the network. Fortunately, perhaps almost miraculously, a theory along these lines now seems to be coming into view.

LAWS OF FORM

THE STUDY OF networks is part of the general area of science known as *complexity theory*. In an abstract sense, any collection of interacting parts—from atoms and molecules to bacteria, pedestrians, traders on a stock market floor, and even nations—represents a kind of substance. Regardless of what it is made of, that substance satisfies certain laws of form, the discovery of which is the aim of complexity theory. Some scientists have disparaged the search for a general science of complexity as a pipe dream, and yet the ideas at the core of this book reveal that there can indeed be sound and specific principles of complexity theory. Some of the deepest truths of our world may turn out to be truths about organization, rather than about what kinds of things make up the world and how those things behave as individuals. The small-worlds idea is one of the newest and most important discoveries in this science of forms, a science that has roots stretching into antiquity.

To the Greek philosopher Plato, a world of perfect forms stood behind all real, tangible objects, and the purpose of all "right thinking" was to come to know those forms, rather than to be led astray by the flawed and imperfect representations of physical reality. The German philosopher Immanuel Kant also saw a deeper reality lurking behind appearances—the reality of the "thing-in-itself," a kind of untouchable essence lying behind all physical objects. In the emerging theory of networks, and in complexity theory more generally, there is an idea that shares a spiritual affinity with these notions, even if it is grounded not in philosophy but in mathematics and empirical science.

For the first time in history, scientists are beginning to learn how to talk meaningfully about the architecture of networks of all kinds, and to perceive important patterns and regularities where they could see none before. This knowledge alone is leading to some remarkable insights. Why does most of the wealth always end up in the hands of the richest few? As we will see, there is a very simple answer to this long-standing puzzle of economics. The answer has little to do with economics and everything to do with the basic workings of networks. Why does the World Wide Web function so efficiently and crash so infrequently? How does the living cell manage to go on living in the face of all kinds of errors and mistakes at the molecular level? Fundamental insights into these questions tumble straight out of the networks perspective, as do some tips on how business firms might organize their management networks so as to take advantage of the same principles of efficient design.

The small-world networks first discovered by Duncan Watts and Steve Strogatz, as well as other kinds of networks that are close relatives, appear to be pervasive in both nature and human society. The World Wide Web now has well over one billion pages, and yet it does not take forever to get from one to another—a few clicks usually suffice, for the very same reason that it takes only six handshakes to go between any two people on our planet. As we will see, there is a kind of innate intelligence in these network structures, almost as if they had been finely crafted and laid out by the hand of some divine architect. Scientists are only beginning to understand where this intelligence comes from, how it can arise quite naturally, and most of all, how we might learn from it.

It is fair to say that the emerging science of networks cannot yet answer all the difficult problems mentioned earlier. How will the loss of one organism affect others within an ecosystem? How can we prevent an economy from lapsing into recession? And why is a human with thirty thousand genes so much more complex than a plant with twenty-five thousand? Puzzles such as these may endure for many years, yet the science of networks at least offers a promising starting point on the road to unraveling them.

In a previous book, *Ubiquity: The Science of History or Why the World Is Simpler Than We Think,* I looked at some other aspects of the emerging science of networks. Recent mathematical ideas point to the possibility that a single scaffolding of logic lies behind tumultuous events of all kinds—from earthquakes and stock market crashes to major military conflicts. Discoveries of the past decade suggest that many of the most important world networks—economies, political systems, ecosystems, and so on—are poised perpetually on the very edge of instability and tumultuous upheaval. As a consequence, it is something akin to a universal law of nature that the course of history must necessarily be punctuated—and quite frequently—by seemingly inexplicable upheavals.

This is a theoretical point about the character of history—that we should expect long periods of relative calm and gradual change to be punctuated by staggeringly overwhelming events that totally reshape the social and political landscape. The point is well illustrated by the terrorist attacks of September 11 and the subsequent events. Little more than a year ago, most people—in the United States, at least—envisioned a prosperous and peaceful future as high technology spurred the economy and democracy spread inexorably over the globe, taking humanity toward something like the End of History. The evening news centered on topics such as the level of consumer spending, the performance of Internet stocks, and the fate of Microsoft. Nothing seemed more remote than global terrorism, anthrax, commando raids, and B-52s. But history, as always, has unfolded in a truly erratic and tumultuous way. History is not mere novelty, as Karl Popper argued, but novelty that arrives unexpectedly in great, terrifying packages.

Since the attacks, we have become accustomed to the idea that the West is battling against a decentralized "network of terrorist cells" that lacks any hierarchical command structure and is distributed throughout the world. This network seems to be a human analogue of the Internet, with an organic structure that makes it extremely difficult to attack. While Osama bin Laden has been the principle focus of U.S. military efforts (as of November 2001, at least), it may well be illusory to see one figure as central. According to George Joffee, a Middle East specialist at the Center for International Studies at Cambridge University in England, bin Laden's group acts more as a "clearinghouse of sorts, providing funds, training and logistical support to other Islamist groups" in countries such as Egypt, Algeria, Somalia, Yemen, Saudi Arabia, and the Philippines.[7] Arresting or killing bin Laden may put important limitations on the network's capabilities, or it may have little effect.

The mantra projected by the White House has been that America faces a "new kind of war" against a nebulous enemy, a phantom network of fiends that is everywhere and nowhere at once. This may be true, and yet the delocalized character of this network is not unique to terrorism. In understanding global politics, it is increasingly important to recognize that the traditional nation-state, dominant for so long, faces threats not only from terrorist organizations but also from transnational corporations that owe their allegiance only to shareholders' profits. By facilitating the global coordination of such network-based forces, computer networks and the Internet are triggering significant changes in the world order.

It would be absurd to suggest that a few discoveries in network theory will enable authorities to move against terrorist networks in any significant way. It is ironic, however, that the terrorists themselves have clearly used our networks against us. On the one hand, the notorious bin Laden was trained and extensively funded by the CIA during the Soviet war in Afghanistan, when he and his compatriots were warmly referred to as "freedom fighters." U.S. taxpayers' funds helped to build the terrorist-training camps that the U.S. military's bombs are now destroying. At another level, other terrorists have struck at the heart of the networks that support the modern world. Nothing connects us together like the postal service, through which letters charged with anthrax have stealthily crept. The initial attacks on September 11 employed our own air transportation network against us, and were organized via remote financial dealings and communications over the Internet. This kind of coordinated effort would have been far more difficult a decade ago.

So networks are in the news and will likely remain there. To understand our world, we need to begin thinking in these terms. *Nexus* focuses on a number of the world's most important networks, as well as some crucial questions: What species are the most crucial to the continued health of the world's ecosystems? What is the best strategy for combatting the spread of AIDS or other diseases? How can businesses exploit the architecture of social networks to improve their ability to gather important information? And how can we best protect the networks on which we all depend, from the telephone system and the electrical grid to the Internet? For these questions and many others, the emerging science of networks offers a deeper perspective on the critical importance of "connections" in our world.

STRANGE CONNECTIONS

The history of mathematics is largely absent from the "culture" of the edu-
cated public, historians and mathematicians included. . . . Like the rainbow,
mathematics may be admired, but—especially among intellectuals—it must
be kept at a distance, away from real life and polite conversation.

—*Ivor Grattan-Guinness*[1]

IN THE SPRING of 1998, the London editorial offices of *Nature* received a manuscript having a somewhat unusual character. *Nature* is the world's premier journal for the latest scientific research on global warming, human genetics, and any number of other areas with broad implications for the future of mankind. But this paper dealt with none of the "usual" topics. It had been sent in by two mathematicians from Cornell University in Ithaca, New York, and yet, lacking almost entirely in equations, it did not much look like a work of mathematics. The only numbers appeared in tables offering data on some peculiar topics, including which actors had played in the same films together over the past half-century.

The manuscript, entitled "Collective Dynamics of 'Small-World' Networks," also contained some curious circular diagrams: rings of dots wired together with curved lines that looked like a pattern you might find in a decorative wallpaper or lace, or in the pages of some thirteenth-century text on alchemy. Nevertheless, the paper was very far from being a hoax. Its topic was serious and grabbed the immediate attention of the journal's editors, who published it a few months later.[2] The two mathematicians, Duncan Watts and Steve Strogatz, had discovered a mathematical explanation for a centuries-old mystery—what we might call the mystery of the small world.

At one time or another, we have all had a small-world experience. On a plane from Denver to New York, you sit next to a man who went to school with your father forty years ago. On holiday in Paris, you strike up a conversation with a woman who turns out to live with the sister of your best friend back in Boise, Idaho. Anyone can produce stories like this. As for myself, I moved a few years ago from the United States to London to take an editorial position with *Nature*. A few weeks after arriving, I went to a party with some new friends. At the party, most people were British, but quite by chance I sat next to a man who had come from the United States just a few years earlier. From where, I asked? Oddly enough, Virginia, the very same state where I had been living. From where in Virginia? Remarkably, Charlottesville, the not-very-large town from which I too had just come. Where had he lived in Charlottesville? Well, as it turned out, on the same street as myself, just a few doors down, even though I had never met him before.

In view of the staggering number of people on Earth, the vast majority of whom have never lived anywhere even remotely close to where I have lived or visited, a chance encounter of this sort seems rather incredible. It was. Since we've all had similar experiences, and more than once, we might wonder, are these strange coincidences trying to tell us something? The social network representing the entirety of humanity is undoubtedly a very large one. According to the United Nations Department of Economics and Social Affairs, the world population topped six billion for the first time on October 12, 1999. Despite the sheer numbers, however, is there some sense in which the world is actually far "smaller" than it appears? Is there something we don't know about that would explain these coincidences?

These questions are what that peculiar paper by the Cornell mathematicians addressed, in three short pages, using no equations and only a few simple diagrams. The answer Watts and Strogatz offered is that the social networks within which we live possess a special and hitherto unsuspected organization and structure that truly make for a small world. This organization is the "something we don't know about," and, in the first part of this book, we will examine their idea in some detail. We also will begin following the story of what it implies not only for our understanding of social networks, but also for the science of networks in general, in biology, computer science, economics, and in our daily lives.

Before turning to the central idea, however, we should look a bit more closely at the motivating puzzle. Coincidences do take place, and a few anecdotes, or even a long list of anecdotes, cannot suffice to show that there are simply too many coincidences to be explained by chance. We should be sure there really is a small-world mystery that needs explaining. Is there any scientific evidence of such a mystery?

LETTERS TO ANYONE

STANLEY MILGRAM BEGAN his famous small-world experiments, mentioned briefly in the prelude, at Harvard University in the mid-1960s. At the time he was a young assistant professor, but he was quickly building a reputation as an immensely creative experimenter. A few years before, Milgram had invented what he called the "lost-letter technique," a method for probing the attitudes of people within a particular community while avoiding the typical problems of social influence and political correctness that often cloud the results of interviews and questionnaires. As a demonstration, he and some graduate students prepared four sets of one hundred letters addressed to Friends of the Nazi Party, Friends of the Communist Party, Medical Research Associates, and a private individual named Mr. Walter Carnap. They then "lost" these letters in places where random people would find them. After a few weeks they discovered that while about 70 percent of the letters to Medical Research Associates and Walter Carnap had been returned, only about 25 percent of the others made it. For probing community attitudes on sensitive issues, this lost-letter technique soon came to be used widely in social psychology.[3]

The small-world experiments were a slight modification of the same method—except that here the point was to learn not about people's attitudes but about the structure of the social network tying them together. With little more than a few envelopes and stamps, Milgram was able to use the U.S. Postal Service as an ally in probing the social structure, and to offer legitimate experimental evidence that the world is, in a social sense, far smaller than we might suspect. Milgram sent out 160 letters, and of those that made it to their final destination, almost all were given to the stockbroker by one of just three of his own friends—as if most of the social routes converging on that man were focused into a few narrow channels. Even more strikingly, almost all of the letters arrived in just six steps, indicating that the social world really is immeasurably smaller than we would expect.[4]

When it comes to experimental design, Milgram was one of the most original social psychologists of the twentieth century. It would be a shame for me to mention his name without taking a moment to describe the one other experiment that gained him lasting fame and made him the subject of controversy throughout his career. In this study, carried out a few years before the small-world study, Milgram asked volunteers to administer, by pushing a button, a sequence of ever-more painful electrical shocks to a man whom they could see strapped into a chair in the laboratory. This man, the volunteers were told, was the "subject" of the experiment, a study aimed at understanding the effects of punishment on learning. In reality, however, the man in the chair was an actor, and the true subjects of the experiment were the volunteers themselves.

Milgram's idea was to see how much suffering ordinary people would be willing to inflict on an innocent man if they were acting under the authority of another, in this case, under the authority of Milgram himself. In the experiment, Milgram asked the man—"the learner"—various questions, and if he failed to give the correct answer, Milgram then asked the volunteer to punish him with a shock. The setting on the electrical voltage generator in the laboratory went from 15 volts, labeled "SLIGHT SHOCK," up to 450 volts, labeled "DANGER—SEVERE SHOCK." The voltage started out at 15 volts and then increased step-by-step at Milgram's request throughout the experiment. At 75 volts, the man began to grunt with the delivery of each shock. At 120 volts, he groaned and complained verbally of the pain, and at 150 volts he demanded to leave the experiment. At still higher voltages, his utterances and complaints grew ever-more desperate, and at 285 volts he responded with "an agonizing scream."

Milgram described the dilemma at the heart of the experiment as follows: "For the subject, the situation is not a game; conflict is intense and obvious. On one hand, the manifest suffering of the learner presses him to quit. On the other, the experimenter, a legitimate authority to whom the subject feels some commitment, enjoins him to continue. . . . To extricate himself from the situation, the subject must make a clear break with authority. The aim of this investigation was to find when and how people would defy authority in the face of a clear moral imperative."[5]

The results were disturbing. At some point during the experiment, most of the volunteers complained that the man was suffering and that the experiment should not go on, and a few finally made the ultimate break with the experimenter. Nevertheless, of forty subjects, twenty-six continued administering the shocks all the way to the 450-volt level.

The results of the experiment ultimately offer little encouragement concerning the ability of ordinary humans to put their moral concerns above their obedience to authority. As Milgram concluded, "Many subjects will obey the experiments no matter how vehement the pleading of the person being shocked, no matter how painful the shocks seem to be, and no matter how much the victim pleads to be let out. . . . It is the extreme willingness of adults to go to almost any lengths on the command of an authority that constitutes the chief finding of the study and the fact most urgently demanding an explanation."[6]

As we will see later in this book, Milgram eventually was able to provide an explanation for this phenomenon, which surprisingly, centers not on the psychology of individual humans but on a pattern of social interaction that seems to emerge inevitably in any functioning social network.

Getting back to the small-world experiment, however, there are a few more details worth mentioning. As it turns out, the results were not quite as definitive as they might appear. Many letters in the experiment simply never got to Milgram's friend, presumably because they were thrown in the trash bin by some apathetic person along the way. Consequently, the letters that did arrive in Boston probably offered a biased picture. If these letters arrived in six steps, it is possible that others were on their way, following paths of ten, twenty, or thirty steps, but ultimately ended their lives in the trash. So Milgram's results may have slightly underestimated the true number of degrees of separation. Perhaps the social world is larger than his results would naively suggest?

THE ORACLE SAYS

IN 1970, IN an attempt to bolster his results, Milgram tried another experiment. It is not surprising that some pairs of people, or even many pairs, are linked by six degrees of separation. The surprise comes if *everyone* is linked by similarly few steps. Milgram reasoned that because of racial segregation in the United States, whites and blacks would be quite distant from one another socially. So, for this experiment, he had the letters start out in the hands of randomly selected whites living in Los Angeles and make their way toward randomly selected blacks living in New York. One might expect this experiment to do a better job of probing the maximum distance within the social network. But when the letters began arriving, the results were no different from the results of the earlier experiment. Again, most letters made it to their destination in about six steps.

Unless you are a real skeptic, it does appear that at least within the United States, the social world is rather surprisingly small. Is the United States somehow special? That would seem unlikely. After all, it is hard to imagine that there is something fundamentally unique in the way Americans make friends and acquaintances that would make the American social network totally unlike that of Switzerland, Brazil, Japan, or any other country. What's more, Milgram's experiments do not provide the only evidence for a small world. A wealth of other indirect evidence points to the likelihood that the small-world property is a general feature of social networks of all kinds.

Six years ago at the University of Virginia, for example, two graduate students in computer science had an idea for a bit of light, personal entertainment. As a lark, Brett Tjaden and Glenn Wasson invented a game they called the Oracle of Kevin Bacon, and put it on the World Wide Web. What is the Oracle of Kevin Bacon? Suppose we think of any two actors or actresses as being connected if they have ever played in the same film together. How many such links does it take to get from, say, Elvis Presley to Kevin Bacon? As in Greek drama, where the Oracle at Delphi answered questions of importance for great heroes, the Oracle of Kevin Bacon responds with great wisdom to questions concerning actors and their relationship to Kevin Bacon. It is next to impossible to stump the oracle. Indeed, it is difficult even to make it break into a sweat.

Type in the name of the contemporary actor Will Smith and the oracle will spit out an answer in a matter of seconds: Will Smith, it reveals, was in *Independence Day* (1996) with Harry Connick, Jr., who was in *My Dog Skip* (2000) with Kevin Bacon. It takes only two steps to go from Smith to Bacon. Try someone from an earlier era, say Bing Crosby, and you get an answer just as quickly: Bing Crosby was in *Say One for Me* (1959) with Robert Wagner, who later played in *Wild Things* (1998) with Kevin Bacon. Again, only two steps. The oracle is unshakeable. Elvis Presley? Elvis Presley was in *Speedway* (1968) with Courtney Brown, who was in *My Dog Skip* (2000) with Kevin Bacon.

When Tjaden and Wasson invented their game, it was only to amuse themselves. "Then we told a couple of friends who had graduated and left school," says Tjaden, "and word got out. . . . Pretty soon more people were using it." Within two weeks, the game was garnering national notoriety, and Tjaden was even flown to Los Angeles to appear on the Discovery Channel. While there, he had the rare fortune of playing the Oracle of Kevin Bacon with Kevin Bacon himself.

What is the oracle's secret? Well, actually, there is no secret, for as it turns out, the world of actors is merely a microcosm of the social world more generally, and so the oracle's task is almost always easy. Bacon has played with some 1,472 other actors and actresses. They are one link away from the man himself. Another 110,315 are separated by two links, and 260,123 by three. The oracle, which finds its answers by sifting through some five hundred thousand names listed in the massive Internet Movie Database,[7] reveals that all but a few hundred actors can be linked to Bacon in six steps or fewer. Indeed, no actor is farther away than about ten links, so no matter what actor's name you type in, the oracle can answer in a snap. The average number of links to Bacon for all actors is only 2.896.

Why does Kevin Bacon occupy such a special position in the acting universe? He doesn't. You can play the Bacon game (or a variant of it called Star Links, also available at the same Web site) using the name of virtually any actor because the entire acting network forms one small world. Choose two actors at random—say, Arnold Palmer and Keanu Reeves—and the computer will link them for you in a short number of steps. In this case, the number of links is only three: Arnold Palmer was in *Call Me Bwana* (1963) with Edie Adams, Adams was in *Up in Smoke* (1978) with Rodney Bingenheimer, and Bingenheimer was in *Mayor of Sunset Strip* (2001) with Keanu Reeves.

Actors are people too, of course, and the actor network is really just a part of the greater social network. So the results of Tjaden and Wasson, based as they are on the statistics for some five hundred thousand actors, lend further credibility to Milgram's findings. As do a few other studies of an even more light-hearted nature. Two years ago, the *New*

York Times had fun playing a game called Six Degrees of Monica Lewin-sky.[8] How many steps separate Monica Lewinsky from, say, the Spice Girls? Only a few, it turns out. The Spice Girls were in the film *Spice World*, in which the actor George Wendt appeared. Wendt also starred on the television show *Cheers* with Ted Danson, who is married to Mary Steenburgen. When Danson and Steenburgen were married, at a site on Martha's Vineyard, President Clinton attended the wedding. And, as we all know, Clinton has been linked with some degree of social intimacy with Monica Lewinsky.

This is all fun and frivolous, and at the same time fully consistent with Milgram's results. While we might expect it to take far more than six links to connect six billion people, it doesn't seem to. So, what is the secret?

NETWORKING

LET'S TRY TO get to the conceptual heart of the puzzle. To visualize the problem, suppose we agree to represent people as dots, and to connect together with a line any two people who are acquainted. To be definite, we agree that people are acquainted if upon seeing one another on the street, they say hello and call each other by name. If we had the where-withal to collect all this information, and a large enough piece of paper, we could do this for the United States or Europe or the entire world. Presumably we would find something that looked pretty messy. But we would find that the overall pattern has an amazing property: starting out from any dot we would be able to travel to any other dot in no more than six steps.

To a mathematician, a bunch of dots connected by lines is known technically as a *graph*. Now, we are all familiar with the word *graph*, but mathematicians use it in a slightly different sense. In the financial pages of the newspaper, we often see graphs of stock prices or of the U.S. gross domestic product over the past twenty years. These types of graphs are consistent with the everyday usage of the word. A graph is something that displays information in a convenient format. In mathe-matics, however, a graph is something just slightly more abstract. It is nothing more or less than a network of dots connected by lines. It need not have any meaning whatsoever. It is simply a logical structure, stripped of any direct connection to the real world.

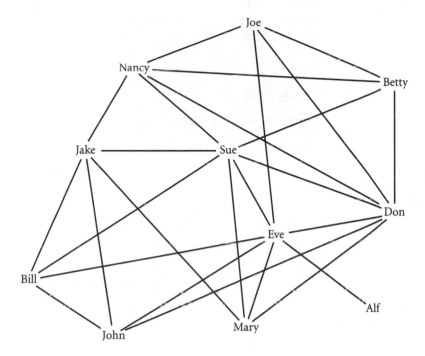

Figure 2. A social network showing the links of acquaintance between fifteen friends.

This understanding of the word *graph* brings us to the basic mathematical puzzle of the small world: Is it possible, and if so, how is it possible, to draw a graph of six billion dots in such a way that one could link any two dots by traveling along just six lines? If someone presented you with a paper with six billion dots on it, how would you begin connecting them so as to give the graph the same character as our social networks?

To get some clues, we might try looking at some real-world friendship networks. Anyone can do this by collecting together a group of friends and then making a graph, representing each friend as a dot and connecting together any pair satisfying our earlier definition of "people who are acquainted." The result would be something like the graph shown in Figure 2, which I have adapted from one that appeared in the magazine *The Futurist* way back in 1975. It shows the links of friendship between various members of a group of fifteen people. In its original version the graph was even messier as it revealed more information such as when each relationship was formed. Do you see any special pat-

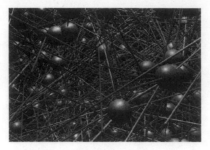

Figure 3. A three-dimensional representation of the social network linking a large group of people in Canberra, Australia. (Reprinted by permission of Alden Klovdahl.)

tern? If so, you are far more perceptive than I. For myself, and I think for most others, this diagram is not particularly revealing or instructive. Some people are connected to one another and some are not, and there is not too much more to say.

Perhaps we need a more sophisticated way of viewing things. For years sociologists have been experimenting with different ways of viewing social networks, and lots of new ideas have become available with modern computers. A few years ago, for example, Australian sociologist Alden Klovdahl used a powerful graphics computer to form a striking three-dimensional image of a network (Figure 3) of friendships among a group of people living in Canberra. This image is rather more sophisticated than that in Figure 2 and has the look and feel of an alien world from *Star Trek*. With this high-tech perspective, we might hope that a pattern emerges, and then we would be able to see at a glance what it is that makes the network special. Again, however, there are no obvious lessons here. Are there any design principles lurking in this mess? Or is the network simply wired up at random?

These images suggest that how a graph works might depend on some fairly subtle features that are not immediately apparent to the eye. In coming chapters we will see what some of these features are, and we will find that despite their subtlety, the way they effect their magic is actually far simpler than any perusal of these diagrams might suggest. We will also see that the mystery of the small world—seemingly a rather inconsequential and light-hearted curiosity—is rather more important than it first appears. In a small world, news and rumors, fashions and gossip can spread more quickly and easily than they would otherwise. So can ideas about what stocks might be worth buying or selling, or about new technologies or strategies for doing business. More ominously, a small world also offers a superconnected web of stepping stones for infectious diseases such as AIDS. The small-world mystery is indeed more than a mere curiosity. It reveals an underlying dynamic of interconnectedness that expresses itself indelibly in who we are, how we think, and how we behave.

THE STRENGTH OF WEAK TIES

Simplicity is the key to effective scientific inquiry.

—*Stanley Milgram*[1]

THE HUNGARIAN PAUL ERDÖS, one of the greatest mathematicians of our era, had no home, no wife or children, and no property or money to speak of. "Some French socialist said that private property was theft," Erdös once proclaimed, "I say that private property is a nuisance."[2] For a full half-century, until his death in 1996 at age eighty-three, Erdös raced around the globe at a reckless pace as a kind of nomadic hobo-genius, coauthoring more than 1,500 papers while sleeping on his colleagues' floors and solving their most difficult problems. Showing up at a mathematician's home, Erdös would declare, "My brain is open," and then work with prodigious energy for several days until his host was too exhausted to continue. Then he was off again, seeking other mathematicians and other problems.

Erdös fuelled his intellect with a steady diet of strong coffee and amphetamines and lived out of a "shabby suitcase and drab orange plastic bag" that he picked up from a department store in Budapest. Mathematics was the focal point of his life. As one of his colleagues, Peter Winkler of AT&T Laboratories, remembered, "Erdös came to my twins' bar mitzvah, notebook in hand. He also brought gifts for my children—he loved kids—and behaved himself very well. But my mother in law tried to throw him out. She thought he was some guy who wandered in off the street, in a rumpled suit, carrying a pad under his arm. It is entirely possible that he proved a theorem or two during the ceremony."[3]

Besides being somewhat eccentric, Erdös was also the Kevin Bacon of mathematics. It is a matter of some pride to mathematicians to talk about their "Erdös number." If you have had the talent and good fortune to have penned a paper with Erdös, then your Erdös number is 1. If you have written a paper not with Erdös himself but with someone else who has written one with Erdös, then you have Erdös number 2, and so on. Remarkably, no mathematician has ever been discovered to have an Erdös number higher than 17, and most—over one hundred thousand—have number 5 or 6. Like all social networks, the collaboration network of mathematicians is a small world.[4]

But this is not why Erdös is of particular relevance to the small-world mystery. Suppose we draw a collection of dots and simply wire them up haphazardly. The result is what mathematicians refer to as *a random graph*. In the late 1950s and early 1960s, Erdös and fellow mathematician Alfréd Rényi wrote a series of classic papers studying such graphs and trying to answer questions about them. As we saw in several examples from the last chapter, the graphs of social networks have no obvious structure or order. To all appearances, they do seem fairly random. So it is natural to wonder, might random graphs offer any insight into the small-world mystery?

As I have said before, it is easy to see the small-world problem as little more than a curiosity, and to suppose that mathematicians could easily make sense of it—if ever they had the time and inclination to do so. The truth is rather different. Graph theory is the branch of mathematics that deals with questions concerning the various ways that a group of things can be connected together, and the theory applies no matter what these "things" might be. So let's take a quick look at what graph theory has to say about the small-world puzzle. Random graphs, in particular, offer an excellent and obvious starting point for the quest to reveal how six billion people can be so intimately linked.

GETTING CONNECTED

IMAGINE YOU HAVE been given the task of building roads to connect up the towns of an undeveloped country. At the moment there are no roads at all, just fifty isolated towns scattered across the map. Linking them together is your task, but it is not quite that simple—you also face some constraints. To begin with, even if you request a road to be built in a precise location, the ever-incompetent Department of Roadways will simply ignore you and put it somewhere else, between a random pair of towns. Ask and it will be built, but there is no telling where.

To make matters worse, the country has very little money, and so you want to build as few roads as possible. The question then is this: how many will be enough? Given unlimited funds, you could command the Department of Roadways to keep building until every last pair of towns were linked together. To link each of the fifty towns to all forty-nine others would take 1,225 roads. That would certainly suffice. But what is the smallest number of roads you need to build to be reasonably sure that drivers can go between any two towns without ever leaving the pavement?

This problem is one of the most famous in graph theory. Of course, it need not involve towns and roads. It could be expressed in terms of houses and telephone links, people and links of acquaintance, a pack of dogs connected randomly by leads, or any of a thousand other things. The essential problem would be the same, and it is by no means an easy one, so you should not worry if you cannot see the answer. In fact, it took the considerable talent of Paul Erdös to solve this puzzle, which he did in 1959. In this particular problem, it turns out, the random placement of about 98 roads is adequate to ensure that the great majority of towns are linked. This may seem like a lot of roads, and yet out of the 1,225 roads that could be built between the fifty towns, 98 represents a mere 8 percent. The Department of Roadways may deserve its bad reputation. But putting links in place at random is not quite as inefficient as it may seem.

Mathematicians are usually less interested in specific problems than they are in generalities, and Erdös was no exception. To a mathematician, mathematics is about a reality that is deeper and more perfect than the physical world, a reality wherein truth can be pursued by logical proof, although the depth of that reality is ultimately inexhaustible. "Mathematics," as Erdös once said, "is the only infinite human activity. It is conceivable that humanity could eventually learn everything in physics or biology. But humanity will certainly never find out mathematics, because the subject is infinite. . . . That is why mathematics is really my only interest."[5]

On this one problem in graph theory, however, Erdös did a pretty thorough job. Indeed, he solved the problem not just for an example with 50 points, but for every example imaginable. He discovered that no matter how many points there might be, a small percentage of randomly placed links is always enough to tie the network together into a more or less completely connected whole. More surprisingly, the percentage required dwindles as the network gets bigger. For a network of 300 points, there are nearly 50,000 possible links that could run between them. But if no more than about 2 percent of these are in place, the network will be completely connected. For 1,000 points, the crucial fraction is less than 1 percent. For 10 million points, it is only 0.0000016.[6]

This perspective already offers one striking insight into the character of the global social network. It is at least conceivable that there might be some pairs of people on Earth who can *never* be connected by any path of intermediate links. Take a potato salesman in Uzbekistan and a laborer on a coffee plantation in Ecuador. Can we really be sure they will be linked, even through a long and convoluted chain of acquaintances? In view of Erdös's mathematics, the answer appears to be yes. After all, for a network of six billion people, the crucial fraction given by Erdös's calculations turns out to be no more than 0.000000004, or about four in a billion.

This number implies that if people were linked more or less at random, the typical person would have to know only about one out of every 250 million people for the entire world population to be linked into a fully connected social web. In total, that works out to only about twenty-four acquaintances for each of us, hardly a large number. With any reasonable definition of *acquaintance,* most of us probably have well over twenty-four. Mathematically speaking, then, it would not be the least bit surprising that any two people in the world can be connected through a pathway of intermediate social links. Erdös's brilliant results come close to proving it.

Unfortunately, this reasoning does not actually deal with the more puzzling aspect of how small the social world seems to be. Just because we are linked into one social whole does not mean that the number of steps between any two people is small. Between Uzbekistan and Ecuador, or between other remote places on Earth, could the number of steps sometimes be as large as 100, 1,000, or 100,000? To find out, we need to ask a different question about these random graphs.

Going back to the sheet of paper with six billion dots, suppose you add lines randomly, connecting first a pair here and then another there, until each dot has a plausible number of "friends," say, fifty or one hundred. Does this lead to a world with six degrees of separation, or to something else instead?

FRIENDS OF FRIENDS

TO START OUT, let's focus on one person—your Aunt Mabel, say, who lives in Florida. And for the sake of argument, let's suppose that each person on the planet has roughly 50 acquaintances. This number would go up or down, depending on how we defined *acquaintance*. But as we will see later, its precise value is not a crucial element of what makes the world so small. So let's stick with 50 to make things simple. On our piece of paper, we put one dot for every person on the planet—six billion in all. Now, Aunt Mabel is one of these people, and she will be connected directly to about 50 others, who in turn will be connected to 50 others. At the level of two degrees of separation, Mabel is already connected to 2,500 of the six billion people. Impressive.

The numbers get more impressive as we go on. Each successive degree of separation increases this number by a factor of 50. At three degrees, there are $50 \times 2,500 = 125,000$ people, and at four degrees, $50 \times 125,000 = 6,250,000$. At five and six degrees, the numbers swell to 312,500,000 and then 15,625,000,000. So at six degrees of separation, we reach a number greater than 15 billion—big enough to include everyone on the planet. The lesson here is that repeated multiplication makes numbers grow quickly. And as a result, it *seems* that the mystery we have been dwelling on is actually no mystery at all. Before we get too sure of ourselves, however, we should think more carefully about what this random graph is saying, for there is something very wrong with our reasoning and with this "obvious" explanation based on it.

Aunt Mabel lives in a house in a small town, goes to church, shops in the local stores, and plays bingo three evenings a week. She leads an eminently normal life. And yet, our reasoning says otherwise. It says that Mabel's social life is rather extraordinary. In a random graph, links pay no respect to the physical proximity or similarity of habits of the people involved. People who work together or who live on the same street are no more likely to be acquainted than are eskimos and aborigines. If we drew this random graph, it would look like a mess of tangled spaghetti and would connect Mabel democratically to people all over the planet. She would have more acquaintances in Russia, China, and India than in her own hometown. This, to put it mildly, is ridiculous.

As social beings, we belong to neighborhoods, companies, schools, villages, and professions. Through work, I know colleagues, and they know not only me but each other as well. Playing bingo, Mabel will have met a number of friends, who will also be friends among themselves. The point is that people are decidedly not wired up at random all over the world. And this simple fact, what we might call the "clustering" of social connections, destroys the calculations we made for the random graph, which now appear as little more than a sterile exercise.

Aunt Mabel may indeed know about 50 people, and each of those 50 may also know another 50. But the bingo friends and the church friends will know many of the same people, as will Mabel's neighbors, or the people she meets when shopping. Each of Mabel's acquaintances will know 50 people, but they probably won't know 50 *different* people— there will be lots of overlaps and repeats. The random graph fails to capture one of the most basic realities of the social world. In a real social network, the numbers will not grow nearly as quickly as our simple-minded calculation said they would.

Perhaps we are making things too complicated. We might return to our piece of paper with its six billion dots and try linking people together in what seems like a more sensible manner, by adding ties from any one person out to the nearest 50 on the page. This way of building a graph captures the idea of community in an obvious way— people tend to know others who live nearby, and within any small area, lots of people will share common acquaintances and friends. If we drew this graph, we would have something resembling more the meshwork of a sophisticated net than a tangled mass of spaghetti. Once we have wired people up this way, we can ask, how many "degrees of separation" do we get? Unfortunately, the answer is not encouraging. Many pairs of people live on roughly opposite sides of the circle. Moving in steps of 50 or even 100, if there are six billion dots, it will take something like 10 million steps to go from one side of the page to the other. This is not even close to being a small world.

So, we face the horns of a dilemma. On the one hand, Erdös's random graphs illustrate how a world of six billion could, in principle, be very small. If our social world were indeed random, then six degrees of separation would be no surprise. But random wiring also wipes out the clusters of local connections that make communities and social groups what they are. Random graphs describe a conceivable world but not the real world. Meanwhile, the orderly graphs, built as they are out of local connections, give rise to a social world with more realistic clusters of friends and acquaintances. But these graphs are not small worlds. So the small-world puzzle endures.

Perhaps there is no simple explanation. Perhaps we are facing some truly weird and spooky phenomenon. The psychologist Carl Jung once speculated about the existence of what he termed the "collective unconscious," a web of archaic psychological connections between people of which we are normally not aware. In Jung's view, these hidden interpersonal connections might account for all manner of strange coincidences, as when a person wakes at 4 A.M. with a start and later discovers that this was the very moment when a loved one died many thousands of miles away. Perhaps this collective unconscious has something to do with all those small-world coincidences. Perhaps these are not coincidences at all but pointers to some hidden level of psychological reality.

This explanation flirts rather closely with the supernatural. But there is a simpler possibility. Could it be that the kind of network needed to depict a social world is neither ordered nor random, but somewhere in-between? Between these two extremes might there be networks having some special organization and unique properties, graphs that would be small worlds but would also offer a realistic account of the clustering that makes our social world what it is? Our social network has not been designed by anyone. It has evolved through countless historical accidents—people meeting people by chance. However random it appears, perhaps some special and rather finely crafted architecture has welled up within it nonetheless.

Sadly, this messy in-between world is an area that graph theorists, including the legendary Paul Erdös, have left almost entirely untouched. Mathematicians have not simply failed to make important progress; rather, they have not even perceived this "in-between" world as being particularly worthy of study. Consequently, there is nothing in three centuries of graph theory that can even give us a hint about how to begin exploring the ill-defined world between chaos and order. Thirty years ago, however, a sociologist set out to examine the most crucial social links that tie our communities together. In so doing, Mark Granovetter never intended to forge his way to the doorstep of an entirely new world of graph theory. Nevertheless, that is precisely what he did.

BRIDGES BETWEEN WORLDS

IN 1973, GRANOVETTER was a young professor at Johns Hopkins University in Baltimore. Stanley Milgram had published his final paper on the small-world phenomenon just three years before, and like many sociologists of the time, Granovetter was fascinated, even amazed, by Milgram's findings. Unlike most, however, he had a hunch about what the secret of the small world might be. In the social networks we have drawn so far, we have not mentioned the strength of the bonds between people. But some bonds are obviously stronger than others. Loosely speaking, we might refer to "strong" ties as those between family members or good friends, or between colleagues who spend a lot of time together, whereas "weak" ties link people who are just acquaintances. Now the mere fact that ties have different strengths would not appear to be pregnant with implications. Excellent scientists, however, often begin with the commonplace, with facts and ideas that everyone knows, and by way of inspired analysis, manage to draw great fountains of insight out of an apparently waterless desert. Granovetter's work offers a prime illustration.

Granovetter noted that if I have two good friends, Paul and Bill, these two people are also likely to be friends with one another. After all, friends tend to have a lot in common. They may live in the same neighborhood, work together, or hang out at the same pub. If two of my friends have things in common with me, they will also tend to have things in common with each other. Moreover, if each spends some time with me, they are also likely to spend some time together—at the very least, when all three of us are together. So if one person is strongly linked to two others, these two also will tend to be strongly linked themselves. There will be exceptions, but generally this should be the case.

In a graph, this principle implies that strong links do not exist in isolation; rather, they tend to fall within triangles. Strong ties between people should almost always appear this way, and triangles with one side missing (Figure 4) should be rare. This idea may seem a bit pointless and technical, but as Granovetter pointed out, it leads to a paradoxical conclusion. Suppose somehow we could remove a strong link from the social network. What effect would this have on the number of degrees of separation? Hardly any. Since strong links almost always appear in special triangles, you would still be able to go from one end

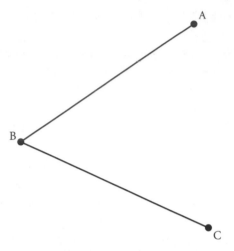

Figure 4. An unlikely social situation. If one individual, B, has strong social links to each of two others, A and C, it is highly likely that A and C will also share a strong link.

of the missing link to the other in just two steps, by moving along the remaining two edges of the triangle. So, you can knock out any strong link without having much effect on "social distances" within the network.

What makes this paradoxical is that you might think that strong social links would be the crucial ones holding a network together. But when it comes to the number of degrees of separation, they aren't; in fact, they are hardly important at all. The crucial links, as Granovetter went on to show, are the weak links between people, especially those that he called social "bridges."

In ordinary language, a bridge is something that provides easy access between two points. The bridge might cross a raging river or a deep gorge, and if it weren't there, it would be much harder to get from one side to the other. In the social context, a bridge has a similar meaning. Mary and Sue might be acquainted, and so in a social graph they would be connected by a single direct link. If that direct link were removed, however, it might suddenly take quite a few steps to connect Mary and Sue through others (Figure 5). In the example I have drawn, adapted from Granovetter's paper, strong links are shown by solid lines and weak links by dashed ones. The number of degrees of separation between Mary and Sue jumps from one to eight if you remove the sin-

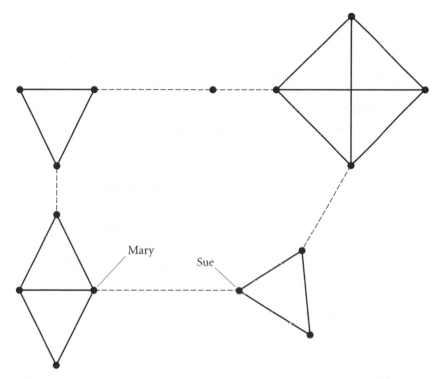

Figure 5. Network diagram to depict the concept of a social "bridge." In this social setting, the link between Mary and Sue is a bridge because its removal has a dramatic effect on the social distance between the two. With the link in place, the distance is just one. If it were missing, it would then take a chain of eight links altogether to go from Mary to Sue.

gle link between them. This link is a social bridge, a crucial connection that binds a portion of the social fabric together.

Granovetter's ultimate point is subtle but extremely important. Because bridges are so effective in tying social networks together, we might suppose that they would be strong links—ties between good friends, for example. But as we have seen, strong links are almost never important in this way. They can be erased without much effect. The truth is just the opposite: bridges *are almost always formed from weak links*. By deftly wielding the knife of elementary logic, Granovetter was able to reach a surprising conclusion: weak links are often of greater importance than strong links because they act as the crucial ties that sew the social network together. These are the social "shortcuts" that if eliminated, would cause the network to fall to pieces. "The Strength of

Weak Ties" is the elegant title of Granovetter's classic paper from 1973.[7] Its central idea sounds strange, but it begins to make intuitive sense when put into the language of the real world.

You have strong links to family members, coworkers, friends, and so on. If your direct link to one of these people were removed, you would probably still be linked by a short path through common friends, other family members, and so on. So no matter how important your relationship is on a personal level, no matter how large a role it may play in your social activity, such a strong link is not likely to be a crucial bridge that glues the social network together. However, you also have acquaintances you rarely see or contact, people you went to college with but even then didn't know very well, and so on. These are your weak links. The guy you worked with during the summer ten years ago, who now lives in Melbourne, Australia, works in the fishing industry and in every way moves in a distinct social world—your link to him may be a social bridge. You may trade letters once every few years. But if this tenuous link were broken, you may well never hear about one another or cross paths again.

Notice that bridging links of this sort do not merely connect you to one other person. They are bridges into distant and otherwise quite alien social worlds. Without that tenuous link to one guy in Melbourne, you might not be connected to anyone there. But because of it, you are linked in two steps to everyone he knows and in three steps to everyone they know, and so on. Strong links do not have this "breaking out" effect; they connect you to people to whom you would be closely connected anyway.

Bridges between worlds, then, have dramatic consequences. And clearly, they must have something to do with what it is that makes the world so small. If six degrees of separation is really true, then six steps suffice to go from any one person to any other. Social shortcuts must play a crucial role in making this possible. To bring the point out more clearly, Granovetter had another clever trick up his sleeve.

NETWORKING

THE TERM *NETWORKING* was one of the buzzwords of the 1980s. If you are searching for a job, the idea goes, or seeking advice on some esoteric question and have little idea which way to turn, your best bet is to exploit the far-reaching fingers of your social network. Put out feelers to friends and acquaintances, and hope that somewhere down the line some gem of information manages to trickle back your way. A friend of your uncle's neighbor may be just the person you are looking for.

Now, if weak bridging links are so important in social networks, you might expect them to be crucial in the networking process as well. If you send out feelers for a job, are you more likely to succeed if you talk to your good friends, or would you do better with weaker acquaintances? As a good sociologist, Granovetter wanted to do more than speculate about the answer, so he devised a clever experiment to find out. He interviewed a number of people who had recently found new jobs through contacts. In each case he asked how they had found the job, and explored the nature of their relationship with the contact person who had been instrumental in making the connection with the employer. One might naively suppose that the strong contacts would be more important. After all, your friends know you better, see you more frequently, and care more about helping you.

Even so, Granovetter found that only 16 percent of the people he interviewed got their jobs through a contact they saw "often," whereas 84 percent got their jobs through contacts they saw "occasionally" or "rarely." The information these people put out into the network—"I am looking for a job"—seems to have spread more effectively and to a greater number of people by moving through weak links rather than strong ones. The explanation seems fairly obvious. Sending out feelers to good friends is certainly easy to do, but the news does not spread very far. Because your friends share common friends, many will soon begin to hear the news a second or third time. But if you broadcast your needs to loose acquaintances, distant relatives you never see, and so on, your news at least has a chance to go farther afield—to escape the confining boundaries of your own social group and get into the minds of a great many people. "From an individual's point of view," Granovetter concluded, "weak ties are an important resource."[8]

A curious study from a few years earlier showed much the same thing. In 1961, social psychologists Anatol Rapoport and W. Horvath visited a junior high school in Michigan, and asked each of the nearly one thousand students to make up a list of his or her eight best friends, putting the best first, then the second best next, and so on. They then used these lists to trace the social connections between the students. Starting with a cluster of ten students, they wrote down for each the

names of their two best friends, then the names of these students' two
best friends, and so on. Moving outward along these strong links,
Rapoport and Horvath eventually produced a map showing the portion of the student population that was connected to the original ten
through a web of strong links. This "strongly connected" cluster
included only a small fraction of the whole school.

Rapoport and Horvath then repeated the procedure, this time using
friends of weaker attachment. They wrote down the lowest two names
on each student's list and let the cluster grow until they had a map of
everyone connected to the original ten through the web of weak links.
This "weakly connected" cluster involved a far higher fraction of the
school.[9] So if the original ten students had started some rumor that
moved only between the best of friends, it would have infected their
own social group but not much more. In contrast, a rumor moving
along weaker links would go much farther. As in the case of people
seeking jobs, information spreading along weak ties has a better chance
to reach a large number of people.

In 1983, Mark Granovetter revisited the topic of social networks
and offered a concise picture of what the distinction between strong
and weak ties implies. He imagined a hypothetical person named Ego,
and considered the structure of Ego's social world:

> Ego will have a collection of close-knit friends, most of whom are in
> touch with one another—a densely-knit "clump" of social structure.
> In addition, Ego will have a collection of acquaintances, few of whom
> know one another. Each of these acquaintances, however, is likely to
> have close friends in his or her own right and therefore to be
> enmeshed in a closely knit clump of social structure, but one different
> from Ego's. The weak tie between Ego and his or her acquaintance,
> therefore, becomes not merely a trivial acquaintance tie, but rather a
> crucial bridge between the two densely knit clumps of close friends.
> . . . These clumps would not, in fact, be connected to one another at
> all were it not for the existence of weak ties.[10]

This is Granovetter's basic insight: the crucial importance in the
social fabric of bridging links between weak acquaintances. Without
weak ties, a community would be fragmented into a number of isolated
cliques.

THE WORLDWIDE REVOLUTION

FOR THOMAS KUHN, an influential historian of science, the essential distinction between revolutionary science and ordinary science is that the former involves a "tradition-shattering" as opposed to a "tradition-preserving" kind of change. In normal scientific work, theories are extended, observations are made more accurate, and understanding grows by a process of accumulation. A scientific revolution, on the other hand, involves throwing out cherished ideas and replacing them with new ones; scientists come to see the world in a different light.

Granovetter's work did not trigger a significant scientific revolution, at least not immediately. For nearly thirty years, his simple but striking insights into the character of social networks and the importance of weak ties—and especially the bridging links that some of them represent—remained virtually unnoticed by other scientists. Meanwhile, Milgram's bizarre findings of six degrees of separation remained unexplained; indeed, few scientists even thought of the small-world problem as something deserving serious consideration. Together, however, these two lines of thinking went a long way toward preparing the ground for a revolution that today is sending out waves through areas as far removed as epidemiology, neuroscience, and economics.

"The first rule of discovery," as the mathematician George Polya once joked, "is to have brains and good luck. The second rule of discovery is to sit tight and wait till you get a bright idea."[11] For scientists as a group, and with respect to the small-world puzzle, it took some thirty years of waiting until someone had that bright idea. It is now time to return to the paper I mentioned in the first chapter, that peculiar manuscript that landed one spring day on the desk of the editors at *Nature*. In it, Cornell researchers Duncan Watts and Steve Strogatz revealed a clever way to connect Granovetter's ideas together with Milgram's, and offered scientists a way to begin forging their way into a conceptual netherworld, a world of complex networks that live between order and chaos.

SMALL WORLDS

This is the remarkable paradox of mathematics: no matter how deter-
minedly its practitioners ignore the world, they consistently produce the
best tools for understanding it.

—John Tierney[1]

IT IS DUSK in the tropical rainforest of Papua, New Guinea. As the
shrieking of parrots and parakeets fades and the tree kangaroos settle in
for a long night, fireflies by the million are taking to the air and lighting
it up like tiny flickering stars. For a while, the fireflies' erratic flashing
will animate the darkening air with a gentle, luminescent chaos. But as
evening turns to night, the chaos will give way to one of nature's most
bizarre displays. Fireflies, first in pairs, then in groups of three, ten, a
hundred, and a thousand, will begin to pulse in near-perfect synchrony.
By midnight, entire trees and clumps of trees will be flashing on and off
with the crisp clarity of neon signs.

"Imagine a tree thirty-five to forty feet high," an eyewitness once
wrote, " . . . apparently with a firefly on every leaf and all the fireflies
flashing in perfect unison at the rate of about three times in two seconds, the tree being in complete darkness between flashes. . . . Imagine
a tenth of a mile of river front with an unbroken line of mangrove trees
with fireflies on every leaf flashing in synchronism, the insects on the
trees at the ends of the line acting in perfect unison with those between.
Then, if one's imagination is sufficiently vivid, he may form some conception of this amazing spectacle."[2] It is more than a spectacle. It is also
a scientific enigma. It is no mystery why individual male fireflies produce flashes of light—they do so to attract mates. Nor is it puzzling why
a group of males might benefit by flashing in unison: as an organized

unit they make a far more brilliant beacon, and so tend to outbid any unsynchronized group in the competition for mates. The puzzle lies with the mechanics. The mental capacity of a typical firefly is not considerable. So how does each insect know in advance when it should, or should not, turn itself on?

On the icy campus of Cornell University in the winter of 1996, this was the question occupying Duncan Watts's thoughts, which may seem just a bit odd given that Watts is a mathematician. Yet Watts was well aware that the fireflies are just one example of a far more widespread phenomenon. Crickets do not send out bursts of light but of sound— chirps produced by rubbing their legs together—and on a hot summer evening a field full of crickets will synchronize their chirping to appreciable effect. Watts was wondering about this phenomenon of fireflies and crickets and also about the odd activity of specialized cells in the human heart. The "cardiac pacemaker" is a bundle of heart cells, each of which sends out electrical signals to the rest of the heart, triggering its contraction. Like the crickets and fireflies, these cells manage to fire strictly in unison, with each collective burst triggering a single heartbeat. If your cardiac pacemaker lost its composure, you would soon be dead.

So how does it work? How can a bunch of cells or crickets or anything else manage to synchronize their activity without the aid of any external guidance or orchestration? This is a question not just of biology but of mathematics, for it seems to reach beyond the details of crickets, heart cells, or what have you, and to point toward some general organizing tendency in nature. Indeed, in the past two decades, neuroscientists have discovered that the synchronized firing of millions of neurons in the human brain appears to be essential to some of the most basic functions of perception. Something quite similar even takes place among the several hundred members of an enthusiastic audience, whose clapping sometimes falls into rhythmic perfection. With a mathematical approach, Watts was hoping to find a way to wrap up a number of these mysteries all at once.

Such a goal may seem a tall order, but fortunately Watts was not starting from scratch. Nearly three centuries ago, the Dutch physicist Christian Huygens was lying in bed one morning when he noticed that the long swinging arms of two grandfather clocks standing against the wall of his bedroom were mysteriously moving in perfect unison. Puzzled, he got out of bed and disturbed one of the clocks. Within a minute, they had again fallen back into rhythm. Huygens then moved one clock to the far side of the room, whereupon the clocks quickly lost their synchrony. Replaced to their original positions, they found it again. The clocks, as Huygens eventually worked out, were influencing one another through subtle vibrations in the floor, thereby drawing themselves to swing in unison.

By 1996, Watts and his doctoral advisor, Steve Strogatz, were almost certain that something similar was going on with the fireflies, crickets, and heart cells. Six years earlier, in fact, Strogatz and mathematician Renato Mirollo had used a computer to simulate a swarm of "virtual fireflies" interacting in much the same way as Huygen's clocks. They supposed that when one firefly lights up, the light influences other fireflies who see it to flash just slightly sooner than they would have otherwise. In the simulations, Mirollo and Strogatz started out with a swarm of fireflies flashing in a completely disorganized way, and found that over time this triggering influence—the flash of one fly inducing others to flash slightly sooner than they would have—made groups of synchronized flashers emerge in the swarm. Each group gathered in more fireflies, and the bigger groups "swallowed up" the smaller ones until the entire swarm was one synchronous flickering whole.[3]

But that was in the computer. The trick of mathematical modeling is to make a model that is simple enough to understand how it works, but detailed enough to relate to reality. And this is where Mirollo and Strogatz ran into trouble. They had been able to prove that a swarm of fireflies affecting one another with their light should almost always manage to synchronize. But for their mathematical reasoning to pay off, they had to assume that when one firefly lights up, its light affects *every other fly*, no matter how close or far away, in exactly the same way, which is a bit crazy. At a crowded football match, you hear the noise from your obnoxious neighbors more than from the fans across the stadium. For crickets or fireflies, the same should be true.

On a mathematical level, this was Watts's real problem: to go beyond this earlier work by figuring out how to "wire up" fireflies in a more realistic way. He could not even be sure that the pattern would really matter. Would a swarm of fireflies synchronize regardless of which fly was "talking" to which? In any event, what was the right pattern? By this time Watts had asked all the firefly experts he could find, but to no avail. "I certainly didn't know," he recalls, "and neither, it seemed, did anybody else."

A DISTRACTING THOUGHT

ONE DAY, WHILE pondering the issue, Watts happened to recall something his father had said to him many years before: "Did you know," he had said, "that you are only six handshakes away from the president of the United States?" At that time, Watts most certainly did not know; in fact, he had never heard of Stanley Milgram or six degrees of separation, and frankly doubted that it could possibly be true. But if it were true, it seemed conceivable that there might be some hidden link to the fireflies. The idea seemed a bit far-fetched, but as he was at a dead end with the fireflies, Watts went to the library and looked up Milgram's original paper anyway. Then he plunged more deeply into the library, in search of anything he could find about the small-world problem.

Over the next few weeks, Watts came upon one rather imposing book on the topic, and a handful of disappointing articles. These publications certainly testified to the reality of the small-world effect but offered little in the way of explanation. Watts also learned that a few imaginative researchers had turned to non-Euclidean geometry, the mathematical framework of Albert Einstein's theory of relativity, in the hope that it might offer some insight into the geometrical character of the social world. Non-Euclidean geometry makes it possible to talk about worlds in which space can be curved and distances work in utterly peculiar ways. Einstein himself famously claimed that in science, "imagination is more important than knowledge," but in this case, as far as Watts could see, the imaginative leap that would connect sociology with non-Euclidean geometry seemed to have landed on infertile ground.

When it came to explaining the small-world phenomenon, Watts could find almost nothing in the literature of three decades, except for Mark Granovetter's incisive but incomplete ideas about the strength of weak ties. Granovetter had argued convincingly that weak social bonds are the most crucial in tying together a society. Somehow, these links ultimately make for a small world. What was lacking was a recipe for building precise mathematical networks that would bring the architecture of such a world into sharp relief. As we have already seen, ordered networks give rise to clusters and cliques just as we find in real social networks. But ordered networks do not have the small-world property—it takes way too many steps to get from one point to another. In contrast, random networks make for small worlds but worlds without clusters, worlds in which there is no such thing as a group of friends or a community.

By this time, Watts had spent more than a month in the library, and was beginning to wonder about himself. None of this had any clear link to the fireflies. And yet he was hooked, which explains why one morning he found himself trudging into Strogatz's office with a stack of papers under his arm and a half-baked proposal in his head. "Why don't we drop the crickets and fireflies for the moment," he said, "and instead try to figure out what it is about the world that puts us all within a few handshakes of each other?" As Watts recalls, he explained what he meant with a mixture of "used-car-salesman enthusiasm and frightened-schoolboy stammering," and fully expected his advisor to laugh him straight out of the office. But to his endless amazement, Strogatz didn't laugh. He liked the idea.

Over the next few hours, Watts filled Strogatz in on everything he had learned, and the two agreed that the proper recipe for a social network somehow would have to involve a peculiar mingling of order and randomness in one network. From a mathematical point of view, the central question was how to do that. Watts also had been thumbing his way through all the classic works of graph theory and could find nothing that might point the way. After a bit of thinking, however, and a bit more scribbling, the two mathematicians finally hit on an idea. It was not a particularly elegant idea—in fact, it looked more like crude engineering than mathematics—but at the very least, it offered a way to get started.

When an electrician enters your home to do some rewiring, he usually arrives with a plan. If an electrician went mad and started slapping wires in at random, the result would be predictably disastrous. Even so, as Watts and Strogatz discovered, a little random rewiring is sometimes beneficial.

A RANDOM PLAN

TO EXPLORE NETWORKS in the netherworld between chaos and order, Watts and Strogatz decided to start out with a fully ordered circular network, each dot being connected to just a few of its closest neighbors. Then they could do some haphazard rewiring. Choosing a pair of dots at random, they could add a new link between them. Then they could

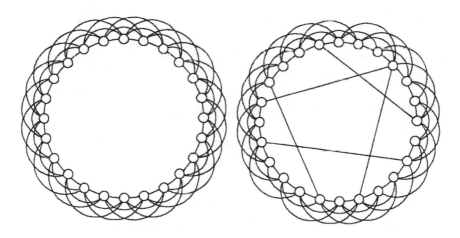

Figure 6. Network evolution. The upper network is fully ordered, with each element being connected to its four nearest neighbors. The lower network comes about through a little re-wiring, with a handful of new links thrown in between pairs of elements chosen entirely at random.

choose another pair and do the same thing, and so on (Figure 6). Suppose the initial circle has 1,000 dots, each connected to its 10 nearest neighbors. This gives, in all, about 5,000 links to start out. To this they could add 10 links at random, and so make a network that remains almost completely ordered but has been seasoned with a splash of randomness. Or they could add in 5,000 new links and get a network that is half random and half ordered. They could choose the mix however they liked.

Using the computer, Watts and Strogatz began doing these kinds of experiments for hundreds of graphs. Each started life as an ordered network, which they then subjected to some random rewiring. As the network's character evolved, they monitored the number of degrees of separation and computed how "clustered" the network was. They already knew what they would find at the two endpoints. The interesting ground was in the middle, and here the computer soon turned up an intriguing surprise.

In their initial network, each point had 10 neighbors, and in principle, a potential total of 45 links could run between these 10 neighbors. In reality, only 2 of every 3 of these points were actually linked, and so the "degree of clustering" was 2/3, or 0.67. This is actually quite a lot, and if you lived in this world, most of your friends would also be friends of one another. The number of degrees of separation, the typical number of steps between points in this ordered world, is also quite large. To travel from one side of the network to the other takes about 50 steps.

Starting with this network, Watts and Strogatz began adding in some disorder. To the original network of roughly 5,000 wires, they had the computer add 50 more at random. The result was a network still almost totally dominated by order, but with roughly 1 percent of its links now placed at random. Not surprisingly, this dusting of disorder hardly altered the clustering of the network at all. The computer calculated a degree of clustering of 0.65, compared to the 0.67 for the original. Not very exciting.

Watts and Strogatz noticed, however, that while the scattered random links were having no effect on the network's clustering, they nevertheless were having a devastating influence on the number of degrees of separation. With no random links at all, this number had been roughly 50; now, with a few random links thrown in, it had suddenly plummeted to about 7. They checked their computer code, expecting to find an error, but could find none. So, over the next few days, they repeated the procedure hundreds of times, trying bigger and smaller graphs, taking the points out of a circle and arranging them in different patterns, and linking them up with a different number of nearest neighbors. It did not seem to matter. No matter what they did, the lightest dusting of random links was always enough to produce a small world.

These middle-world networks were pulling off a curious trick: managing to be both small worlds and highly clustered at once. A purely random network, as we have learned, has the small-world property. But for 1,000 people linked together randomly, the degree of clustering turns out to be about 0.01,[4] which is not even close to what one finds in a real social network. These "small-world" graphs, as Watts and Strogatz were soon calling them, were mingling the best of both worlds. And with the considerable aid of hindsight, it is easy to see how they were doing it.

Suppose you live in an ordered network and want to travel between two distant points, A and B. Unfortunately, this inevitably means embarking on a grueling step-by-step march. After all, the links in an ordered network only connect points that are close to one another;

there are no shortcuts or bridges between distant points. Throw in some random links, however, and your network would be transformed. A few of the new links, just by chance, will probably stretch between points that are far apart. Now if you face a long journey, you can hitch a ride on some long-distance expressway, thereby taking the slog out of the journey, and then follow up with a few more short steps to reach your precise destination.

So let's go back to a circle of six billion, the world's population, with each person linked to his or her nearest 50 neighbors. In the ordered network, the number of degrees of separation is something like 60 million—this being the number of steps it takes, even moving 50 at a time, to go halfway round the circle. Throw in a few random links, however, and this number comes crashing down. According to Watts and Strogatz's calculations, even if the fraction of new random links is only about 2 out of 10,000, the number of degrees of separation drops from 60 million to about 8; if the fraction is 3 out of 10,000, it falls to 5. Meanwhile, the random links, being so few in number, have no noticeable effect on the degree of local clustering that makes social networks what they are.

These small-world networks work magic. From a conceptual point of view, they reveal how it is possible to wire up a social world so as to get only six degrees of separation, while still permitting the richly clustered and intertwined social structure of groups and communities that we see in the real world. Even a tiny fraction of weak links—long-distance bridges within the social world—has an immense influence on the number of degrees of separation. What's more, we find here an explanation not only for why the world is small, but also for why we are continually surprised by it. After all, the long-distance social shortcuts that make the world small are mostly invisible in our ordinary social lives. We can only see as far as those to whom we are directly linked—by strong or weak ties alike. We do not know all the people our friends know, let alone the friends and acquaintances of those people. It stands to reason that the shortcuts of the social world lie mostly beyond our vision, and only come into our vision when we stumble over their startling consequences.

This clever architecture seems to wrap up the small-world mystery once and for all. And yet for Watts and Strogatz, this was hardly the end of the discovery—in fact, it was only the beginning.

INTERCONNECTIONS

WITH THEIR SMALL-WORLD graphs, Watts and Strogatz had little doubt they were on the right track. But to check their findings out carefully, they turned to the Internet Movie Database, the biggest set of data they could find for a social network. If small-world graphs really capture the character of actual social networks, then the network of actors linked by having played together in some film should have roughly the same number of degrees of separation as a random network of the same size. At the same time, however, it should also reveal a far higher degree of clustering than a random graph, being nearly as highly clustered as a perfectly ordered graph. For the real acting network, Watts and Strogatz let the computer calculate the numbers, finding 3.65 for the number of degrees of separation and 0.79 for the degree of clustering.

According to the database, each of the 225,226 actors considered has on average 61 links to others with whom they have played at one time or another. So for comparison, Watts and Strogatz set up a random network of the same features. (Without the numbers being the same, a comparison would be meaningless.) For the random network, the computer calculated the number of degrees of separation to be 2.99, and the clustering to be 0.00027. As suspected, the acting world is both highly clustered and a small world all at once. It is a small-world network.

These results were satisfying, to say the least. But with their computer program up and running, Watts and Strogatz were not content to stop there. The recipe for their small-world networks was deliciously and surprisingly simple. There was nothing in the recipe that referred to people per se, or to social connections in particular; rather, the recipe reflected a very basic scheme in graph theory. Hence, it was natural for them to wonder if some other real-world networks might reveal a similar architecture. By chance, Watts had recently come across a book about the structure and history of the U.S. electrical power grid, and this seemed an obvious place to start.

The overall power grid is actually made of three subnetworks, known as *interconnections*. One supplies power to the eastern part of the United States; another, to the region west of the Rocky Mountains; and the third supplies power solely to the state of Texas. Given their size and complexity, it goes without saying that these networks were not designed by any single planner or planning group. Indeed, they reflect the legacy of a thousand historical accidents as new technologies were invented and as new generators and transmission lines were thrown in place to meet the needs of industry and a growing population. On a map, the power grid looks like a sprawling tangle of totally disorganized lines. On a hunch, however, Watts and Strogatz compiled data for nearly five thousand generators, transformers, and substations in the American West, as well as the disorderly web of high-voltage transmission lines connecting them, and fed everything into the computer.

On average, all over the West, each element is linked to roughly three others. What fraction of these elements are themselves linked together? Again, this fraction represents the degree of clustering, and for the power network the computer revealed that it is ten times higher than it would be if the network were built at random. At the same time, the average number of transmission lines that you need to follow to get from one element to another is only 18.7. This is not six degrees of separation, but it is not too far away. Indeed, if the network had been wired totally at random, with five thousand elements and three links on average coming from each, there would be twelve degrees of separation. Again, the pattern is the same: the real-world network is highly clustered and a small world at once.

At first glance, this seems rather bizarre. The links that tie our societies together have grown out of the confluence of all the various social forces, and through the activities of families, schools, businesses, clubs, and so on. None of this would seem to have much in common with the manifold forces of economics, technology, and population growth that have shaped the architecture of the power networks on which we depend. And yet power networks appear to possess the same basic wiring plan as our social networks. If both kinds of networks had turned out to be totally random, and similar for that reason, this would be no surprise. But the small-world architecture is far from random.

Getting to the bottom of this bizarre coincidence was certainly near the top of Watts and Strogatz's list of things to do. For the moment, however, that task would have to wait. It was time to take another look at the fireflies.

HOW TO WIRE A WORM

IN THEIR EARLIER stab at the synchronization puzzle, Mirollo and Strogatz had assumed that each firefly would "see" and respond to every other one. In a swarm of 10,000 fireflies, this would make for a total of about 50 million communication links between fireflies (one link for each conceivable pair). With such a dense web of connections, the group would have little trouble achieving and maintaining global organization, which perhaps is not surprising. Armed with the small-world idea, Watts and Strogatz decided to revisit the issue. Could a swarm of fireflies achieve the same result with fewer communication links?

It would seem a bit more realistic to suppose that each firefly responds mostly to the flashing of a few of its nearest neighbors, although a rare few might also feel the influence of a fly or two at a longer distance. A few flies might have a particularly brilliant flash, and so be visible to others far away, or a few genetic oddballs might respond more to fainter flashes than to bright ones. In either case, the flies would interact with one another in something like the small-world pattern. With this architecture, the same 10,000 flies would be trying to make due with many thousands of times fewer communication links than before. In the old pattern, after all, each fly had influenced roughly 10,000 others. Now each would affect only a handful, perhaps as few as 4 or 5. This change in wiring diagram amounts to hacking out more than 99.9 percent of the links between flies. You would hardly expect any communication network to survive such a savaging, and yet this one did.

In a series of computer simulations, Watts and Strogatz found that the insects were able to manage the synchronization almost as readily as if everyone were talking to everyone else. By itself, the small-world architecture offered a reduction in the required number of links by a factor of thousands. There is a profound message lurking here—a message not about biology but about computation.

From an abstract point of view, a group of fireflies trying to synchronize themselves is making an effort in computation. As a whole, the group attempts to process and manage myriad signals, countersignals, and counter-countersignals traveling between individual fireflies, all in an effort to maintain the global order. This computational task is every bit as real as those taking place in a desktop computer or in the neural network of the human brain. Whatever the setting, computation requires information to be moved about between different places. And since the number of degrees of separation reflects the typical time needed to shuttle information from place to place, the small-world architecture makes for computational power and speed.

Of course, no one knows how fireflies are really "connected" within a swarm. Indeed, only a few species in Malaysia, New Guinea, Borneo, and Thailand manage the synchronization trick. Watts and Strogatz had not answered all the questions about the fireflies, and many remain unanswered still.[5] Nevertheless, they had learned that in terms of computational design, small-world architecture is an especially important "good trick." When it comes to computation, though, nothing is so wondrous as the human brain. And so it was natural to wonder, might the brain also exploit the small-world trick? Neither Watts nor Strogatz knew enough neuroscience even to hazard a guess, but fortunately, Watts returned to the library again—this time to learn some biology.

As it turns out, unraveling the tangle of connections among the several hundred billion neurons in the human nervous system is still beyond scientific capabilities. But in the 1960s, Sidney Brenner and colleagues from the Salk Institute in California took a small step along the way. At the time, Brenner was convinced that most of the fundamental questions of biology had already been answered, and was determined to turn his attention to the next level: to understanding the workings of complex biological networks. In this spirit, he and his colleagues took on the task of mapping out the network of connections among the neurons of the nematode worm *Caenorhabditis elegans*. About 1 millimeter in length, *C. elegans* thrives in decaying vegetation all over the world and has only 979 cells when fully developed. In contrast to the staggering complexity of the human nervous system, *C. elegans* has only 282 neurons.

Like the Human Genome Project, mapping the network of connections in the worm took a full decade, but when Brenner and his colleagues finished, biologists found themselves with the first complete map of an entire nervous system. Watts and Strogatz got hold of the map that Brenner's group had produced, and subjected it to the now-familiar architectural analysis. On average, each of the worm's 282 neurons are linked directly to about 14 others. Using a computer to calculate the degree of clustering, Watts and Strogatz found a high number—0.28. In a random graph with 282 elements and 14 links coming from each, the degree of clustering should be no more than 0.05. But while the neural network of *C. elegans* is four times as clustered as a random network, the number of degrees of separation is only 2.65. Compare that value with the value for a random network, 2.25.

Once again, this raises as many questions as it settles. Like the power grid, a rudimentary worm has also discovered the small-world trick. How? Is this simply a weird coincidence? Or does it point to some deeper design principle of nature?

In their three-page paper in the June 4, 1998, issue of *Nature*, Duncan Watts and Steve Strogatz unleashed all these findings on an unsuspecting scientific community.[6] Their paper touched off a storm of further work across many fields of science. As we will see in upcoming chapters, a version of their small-world geometry appears to lie behind the structure of crucial proteins in our bodies, the food webs of our ecosystems, and even the grammar and structure of the language we use. It is the architectural secret of the Internet and despite its apparent simplicity is in all ways a new geometrical and architectural idea of immense importance.

BRAIN WORKS

A theory has only the alternatives of being wrong or right. A model has a
third possibility: it may be right but irrelevant.

                                                    —*Manfred Eigen*[1]

IN THE FINAL decade of the eighteenth century, a brilliant Viennese
physician named Franz Joseph Gall proposed a radical new theory of
the brain. At the time, the human mind was believed to be the seat
of the immortal soul, and examining its deeper nature was the domain
of philosophers. Immanuel Kant had proclaimed space and time to be
the natural and irreducible categories of the mind, basic preconditions
of the way it filters and perceives reality. But on the matter of the physical brain, what the three and a half pounds of tissue is made of and
how it works, scientists knew next to nothing. Gall himself was totally
ignorant of the brain's hundred billion neurons, hundreds of trillions
of interconnections, and more than a thousand kilometers of cabling.
Even so, he had made an astonishing and revolutionary discovery—or
so he thought.

In his work as a physician, Gall had encountered patients with all
sorts of peculiar personalities. Some were notably selfless and kind;
others, ruthless and ambitious; and still others were strikingly intelligent and blessed with mathematical or poetic talent. Over more than a
decade, Gall tirelessly recorded the characteristic traits of his patients,
while quietly amassing data on the sizes and shapes of their heads. He
collected hundreds of human and animal skulls, manufactured countless wax molds of brains, and put calipers to the foreheads of friends
and pupils. On the basis of his observations, he finally came to the con-

clusion that the brain, like the body, possesses distinct organs of various kinds.

As Gall argued, "The same mind which sees through the organ of sight, and which smells through the olfactory organ, learns by heart through the organ of memory, and does good through the organ of benevolence."[2] Of course, if he was right and these organs of the brain were real, it should be possible to locate and measure them, and Gall insisted that this was indeed the case. To seek the organs in someone's head, Gall recommended running the palms over the surface of the scalp, feeling for any unusual bumps or depressions. The idea was that any larger, overdeveloped organ would push the skull outward, causing a protuberance; an underdeveloped organ, on the other hand, would leave an indentation. By seeking the significant bumps and depressions, Gall insisted, one could learn which faculties were overdeveloped or underdeveloped in any individual, and get a quick read of their character.

To make the reading of heads easier, Gall even mapped out the positions of the various organs and listed the twenty-seven faculties with which they were associated. These included parental love, friendly attachment, ambition, and the sense of cunning. Other organs, he claimed, were the seats of ability for music or arithmetic, or for mechanical skills or poetic talent. Using Gall's map, a skilled phrenologist, as practitioners of the technique soon became known, could assess a personality with fingers and palms in just a few minutes. Gall was invited to lecture at major European universities and to demonstrate his method to kings, queens, and statesmen. (He allegedly annoyed Napolean by detecting in the contours of his skull a distinct lack of philosophical talent.)

Gall's only regret was that he did not have more skulls to study. As he wrote in a letter to a colleague, "It would be very agreeable to me, if persons would send me the heads of animals, of which they have observed well the characters; for example, of a dog, who would eat only what he had stolen, or one who could find his master at a great distance. . . ."[3]

Today, of course, the science of phrenology has been completely discredited. No careful scientific study has ever discovered a legitimate link between personality and the shape of the head, and Gall appears to have been fooling himself in thinking that he had found one. Nevertheless, the Viennese physician did get some things right; in fact, his ideas initiated a way of thinking that takes a central position in scientists' picture of the brain.

No one before Gall had conceived of the brain as an assembly of distinct modules, each responsible for a different task—speech, vision, emotions, language, and so on. But in the modern neuroscience laboratory, researchers can see these different modules in action. Functional magnetic resonance imaging is a technique that uses radio waves to probe the pattern of blood flow in the brain, revealing how much oxygen its various parts are using at any moment. This in turn reflects the level of neural activity. With this method, researchers can watch as different regions "light up" on a computer screen as a subject deals with different tasks, such as responding to a verbal command or recognizing a taste. When a subject is given a new telephone number to remember, the hippocampus becomes active, this being the brain region heavily involved in the formation of new memories. Other regions in the brain control hearing and vision, or basic drives such as aggression and hunger. These are not quite the organs Gall had in mind, but they are significant and distinct processing centers within the brain.

Gall was also the first to focus the attention of neuroscientists on the special region of the brain where these centers reside: the thin, gray outer layer known as the *cerebral cortex*. For centuries, the cortex had been thought of as an unimportant protective layer. In actuality, though never more than a few millimeters thick, this layer contains most of the brain's precious neurons. The surface of the cortex is smooth in small mammals and other lower organisms, but it becomes highly convoluted and folded in creatures roughly bigger than mice, this being necessary to fit the bigger brains inside the skull. If you could stretch the cortex of the human brain out flat, it would cover the surface of a picnic table. This intricately folded and delicately packed cortex is where higher intelligence resides. It is the part of the brain that lets us speak, make plans, learn calculus, and invent excuses for being late.

In short, the cortex is what makes us distinctively human. And it is indeed organized, as Gall suspected, into something like a set of organs. Of course, the brain is hardly an anarchy of modules working in blind independence of one another. These modules have to communicate in order to coordinate overall brain activity. When speaking, we not only choose the right words and put them together properly, but also access memories and control the timing of our speech and its emotional tone.

We may recall a name, whisper it to a friend, and gesture with our hands to some additional effect. These actions involve many of the brain's functional regions working in effortless combination, with information shuttling rapidly and efficiently between them. All this activity raises an obvious question: what wiring pattern does the brain use to provide this efficiency?

A THOUGHTFUL ARCHITECTURE

A REGION OF the human brain no larger than a marble contains as many neurons as there are people in the United States. In the crudest picture, each neuron is a single cell with a central body from which issue numerous fibers. The shortest of these, known as *dendrites*, are the neuron's receiving channels, while the longer fibers, known as *axons*, are its transmission lines. The axons running away from any neuron eventually link up with the dendrites of other neurons, providing communicating links. Details on the full structure of any neuron would require pages, perhaps a book, but these features are the essentials.

By a long shot, most of the neurons link up with others nearby, within the same functional region, for example, be it the hippocampus or Broca's area, a part of the brain involved in the production of speech, as the French neurologist Paul Broca discovered in 1862. Some axons run a bit farther and link up with neurons in neighboring brain regions. Most neighboring regions are linked in this way. From a whole-brain perspective, if we think of the various functional regions of the brain as the nodes of a network, these "local" connections sew the brain together into one connected whole, not unlike an ordered network. However, the brain also has a smaller number of truly long-distance axons that link brain regions that lie far apart, sometimes even on opposite sides of the brain. Consequently, we have many local links and a few long-distance links, something that begins to sound like the small-world pattern. As researchers have recently found, it is relatively easy to bring this pattern into sharper focus.

At the University of Newcastle in England, psychologist Jack Scannell has spent more than a decade mapping out the connections between different regions of the cerebral cortex. As it happens, Scannell's studies have focussed on the brains of cats and monkeys rather than humans, and yet given the great similarities between all mammalian brains, the results almost certainly apply to human brains as well.[4] Earlier research in the cat identified some fifty-five regions of the cerebral cortex, each associated with a distinctive function; in the macaque, the number is sixty-nine. In these brains, there are roughly four to five hundred significant links connecting the different regions, links formed not only by single axons but also by more appreciable streams of many axons running in parallel.

To find out how these links are arranged, Vito Latora of the University of Paris and Massimo Marchiori of the Massachusetts Institute of Technology used Scannell's maps. Analyzing these networks in the terms set out by Watts and Strogatz, they found the signatures of a strikingly efficient network architecture.[5] In the cat brain, for example, the number of degrees of separation turns out to be between only two and three. The number is identical in the macaque brain. At the same time, Latora and Marchiori found that each of these neural networks is highly clustered. In other words, it seems that what is true for good friends is also true for regions of the cerebral cortex: if one brain region has links to two others, then these two other regions are also likely to share a link.

From a functional point of view, these features make obvious biological good sense. If you mistakenly pick up a burning stick, sensory neurons immediately send signals racing toward the brain. These signals trigger a chain reaction of neurons stimulating other neurons, which eventually reaches motor neurons that send signals back to the fingers, vocal chords, and muscles of the tongue and mouth—you drop the stick and cry out in pain. If transmitting the information involved hundreds or thousands of steps between neurons, reflex responses would take far longer than they do. The small-world pattern guarantees that the brain's diverse functional parts reside only a few steps from one another, binding the entire network into one intimate unit.

Quick and efficient transmission of signals is the simplest and most obvious benefit conferred by the small-world structure. But there is another benefit. In a social network, as Mark Granovetter pointed out, the clustering of links among a group of good friends implies that if a few of them were removed from the network, the others would still remain closely linked. In a clustered network, in other words, the loss of one element will not trigger a dramatic fragmentation of the network into disconnected parts. In the brain, this organization may also play a useful role, for the damage or destruction of one particular region would have little effect on the ability of signals to move among and coordinate the other regions. Patients with damage to Broca's region, for example, cannot understand speech, yet they hear perfectly well, do mathematics, and make plans for the future with no difficulty. If damage to this area also severed communications between, say, the visual cortex and the hippocampus, or at least made signals travel long distances to get from one region to the other, then short-term memory of visual information might also be impaired. The small-world architecture seems to prevent this. It not only makes the brain efficient and quick but also gives it the ability to stand up in the face of faults.

The mammalian brain, human or otherwise, is far more than a device for generating efficient reflex responses or hanging together under adversity. The small-world network in your head appears to work magic in many other ways as well.

CONSCIOUS COMPLEXITY

THE NATURE OF consciousness remains one of the most perplexing of all scientific mysteries. The human brain is a lump of material made of ordinary cells working on chemical and electrical principles. From the point of view of fundamental biology and the laws of physics, it is a perfectly ordinary physical machine, although one of undeniable complexity. But if the brain is merely physical stuff doing ordinary physical things, what is the seat of the seemingly spiritual entity that can perceive itself and say, "I"? Are our emotions and sense of responsibility merely the consequences of the mechanical laws of physics when combined in a setting of sufficient complexity? Or is there some mysterious extra ingredient in the brain's workings, without which consciousness would be impossible?

Philosophers, psychologists, computer scientists, and neuroscientists are still arguing over these questions. No one can say for sure what consciousness really is, point to the exact neurons from which it issues, or explain how we might create it artificially. And yet neuroscientists have taken impressive steps in exploring the neural activity associated with consciousness, as well as some of the mechanisms by which the conscious brain seems to work.

The amazing power of the brain arises in part from its ability to respond to the external world with an immense repertoire of possible conscious states. Suppose you look through a window and see a person approaching your house. In a fraction of a second, your brain will flip through thousands of different conscious states, each with a slightly different awareness of the person's ever-changing position. With the addition of emotions, expectations of the future, awareness of sounds, links to memories, and so on, the set of possible conscious states is clearly enormous. Yet your brain at every instant quickly settles into just one of these innumerable states, chosen in delicate correspondence with the external world and your own personal history and condition.

This extreme flexibility is what makes us so complex and gives us great adaptability in the face of a changing world. Equally impressive, however, is the depth of the brain's conscious organization. To be conscious of someone approaching the house is to have a visual image, to appreciate the aspect of movement, to place the image in the context of a particular window through which it is seen, and to link the image to memories of people or situations possibly connected to this person. The brain binds these diverse aspects and countless others into a single, indivisible mental scene, which loses its meaning if it is broken down into its components. To put it another way, the brain acts as a remarkably well-coordinated unit to produce just one completely integrated conscious response at any instant.

What goes on with the neurons to make all this take place? On the one hand, the sheer number of neurons in the brain may acount for the broad range of its possible states. It is less easy, however, to understand what neurons have to do to tie together the different components of a conscious scene. Neuroscientists are probably years away from understanding this in detail, and yet they have uncovered some significant clues. Researchers have learned, for example, that consciousness always involves the activation of neurons from many regions of the brain—it seems to depend on their coherent engagement into one overall pattern. And the mechanism of this engagement, at least in part, is neural synchrony.

In a striking experiment in 1999, for example, neuroscientist Wolf Singer and colleagues from the Max Planck Institute for Brain Research in Frankfurt, Germany, devised a way to present a cat with two distinct series of stripes moving at right angles to one another. The experimenters could adjust the brightness of the two patterns, and in this way control what the cat perceived. If one set of stripes were brighter than the other, the cat would see the two sets as independent features; if the brightness were equal, however, the cat instead would see the stripes melded together as if there was only one pattern, a single checkerboard pattern moving in a third direction (halfway between the directions of movement of the two sets of stripes).

This clever setup enabled Singer's team to study how neurons in the cat's brain respond as the cat goes from seeing the stripes as distinct and unconnected to seeing them bound together into a conscious whole. Slowly altering the brightness, while monitoring the activity of more than a hundred neurons over a wide area in the cat's visual cortex, they found that when the cat was seeing two distinct sets of stripes, two corresponding sets of neurons were firing. Notably, they were out of synch with one another. However, when the brightness was adjusted so that the cat perceived just one pattern, the two sets of neurons fell into close synchrony.[6] The synchronous firing seemed to bind the two distinct features together into one conscious element.

This experiment represents the state of the art in brain research, as the team had the ability to record activity from over a hundred neurons at the same time. At the California Institute of Technology, experimental neuroscientist Gilles Laurent and his colleagues used a similar technique in studying the brains of locusts, and here too they discovered an important role for neural synchronization.[7]

In the locust, the olfactory antennal lobe is a group of about eight hundred neurons that takes information from the olfactory "smell" receptors and relays it toward higher regions of the brain. When a locust smells an interesting odor, these neurons respond very quickly by firing in a synchronized pattern at about twenty times a second. This is only part of the organizational response, however. Relative to the collective organized firing of the group, each neuron also maintains its own specific timing, just slightly ahead or behind the average. These findings imply that the neurons are storing information not only at the group level, by virtue of their synchrony, but also at the individual level, in their exact timing. So lots of information gets sent upward for further processing.

Again, as in the cat, synchrony seems to be central to the way the network of neurons accomplishes its function. And it seems sensible to wonder if the small-world architecture of the nervous system might not be crucial in allowing this synchrony to take place. Think again of the fireflies and crickets. As Watts and Strogatz discovered, the small-world pattern of links would offer a great benefit to a collection of fireflies trying to synchronize their firing. In neurons, it is indeed beginning to appear that the small-world trick is not only a good idea but one of the most basic prerequisites for the brain's fundamental functions.

THINKING QUICKLY

WOULD THE SMALL-WORLD trick really help a network of neurons to synchronize? Would it offer any other advantages or disadvantages as well? In 1999, Luis Lago-Fernández and colleagues from the Autonomous University of Madrid tried to find out by studying networks of neurons in much the way Watts and Strogatz had studied collections of fireflies. Specifically, they created a virtual model of the locust's olfactory antennal lobe and put it through its paces to see how it would respond to a stimulus.

No one has ever studied the layout of neurons in the locust closely enough to know its real architecture. So Lago-Fernández and his colleagues tried out several possibilities. To begin with, they wired the neurons together as in a regular or ordered network. To make the simulations realistic, they used detailed models for the behavior of each of eight hundred neurons, models that were developed over half a century by the painstaking efforts of experimental neuroscientists. Using the computer, the team could apply a stimulus to a small fraction of neurons in the network and then monitor the network as the activity spread throughout.

Hitting this network with a triggering impulse—the analog of the locust detecting a significant smell—the team found that this regular architecture offered a decidely inadequate response. To be sure, the various neurons fell into synchrony and were able to generate the coherent response required to register the smell, much as happens in any living locust. There was one serious problem, however. After presentation of the stimulus, the neurons took a long time to manage their synchrony, far longer than neurons in the locust. If a locust's brain were wired this way, it would be a sluggish performer.

Next the researchers tried a random wiring pattern. But this did not work much better. The few neurons triggered by the initial odor quickly stimulated others, and the activity raced rapidly throughout the entire network. This is not surprising, since the number of degrees of separation is so much smaller than for the ordered network. There was another problem, however. In this random network, the neurons were never able to synchronize their activity. The network responded quickly but ultimately in a disorganized way, failing to record the detected smell in coherent neuronal oscillations.

Like many scientists, Lago-Fernández and his colleagues by this time had read about Watts and Strogatz's work on small-world networks, and they decided to give these a try. Going back to their ordered networks, they added in a few long-distance links between the neurons and then tested the resulting small-world networks in further simulations. The results seemed almost too good to be true. The small-world network responded quickly to the stimulus, much like the random network, and yet the neurons also fell into a synchronized firing pattern, thereby establishing the coherence that in the locust plays a crucial role in representing the perception. Again, the small-world pattern seems to offer the best of both worlds—suggesting, indirectly, that this is how the locust's nervous system is wired up.

As a bonus, Lago-Fernández and his colleagues also discovered one other impressive effect. One of the more subtle characteristics of the neural response in the real locust is the precise timing of each neuron with respect to the group oscillation. This precision enables the network to store far more information than it could otherwise, information that perhaps serves to distinguish different but closely related smells or to represent other features such as the strength of the odor. In their simulations, the team found that the small-world networks were the only ones capable of synchronizing quickly while also permitting each specific neuron to adopt a specific timing with respect to the overall oscillation. What this really means in the context of neuroscience is hard to say, but it appears that the small-world architecture permits the networks not only to respond quickly and coherently but also to store information compactly.[8]

None of these findings really explain how the brain works or what consciousness is. However, they do offer some clues concerning how the brain achieves the organized activity that underlies consciousness. Synchronous neural activity appears to play a central role in conscious functioning, and the small-world architecture facilitates the process. The wiring diagram of the brain may look like a mess, and yet it most certainly is not—as a small world, there is actually a great deal of sense lurking within it.

NATURAL PLANS

ABOVE THE DOOR to Plato's academy in ancient Athens was an inscription that read, "Let no one enter who does not know geometry." For Plato, as for the followers of Pythagoras some nine centuries earlier, geometry was not simply a means toward practical ends. Its spiritual promise lay precisely in its refined pureness and abstraction away from all practical concern. According to Plato, if a man were to contemplate the absolute truths of geometrical reality, in so doing he would touch near the heart of the universe, brushing against a reality that is deeper than the reality we usually know. In this sense, geometry was a spiritual enterprise meant in part to better the man and to educate the soul toward perfection.

Or as Socrates put it in Plato's *Republic,* "The man whose mind is truly fixed on external realities has no leisure to turn his gaze downwards upon the petty affairs of man . . . but he fixes his gaze upon the things of the eternal and unchanging order, and seeing that they neither wrong nor are wronged by one another, but all abide in harmony as reason bids, he will endeavor to imitate them and, as far as may be, to fashion himself in their likeness and assimilate himself to them. . . . "[9]

In this task, the raw materials for pondering geometry were to be found in Euclid's *Elements,* written about 300 B.C. Euclid sought the perfect forms of geometrical truth by starting from simple and apparently "self-evident" facts or axioms, and using the divine light of reason to pull from them theorems of ever-greater complexity. For Plato, even God himself seems to have followed the Euclidean precepts in assembling the universe. Plato speculated about the origins of the universe and imagined God using the building blocks that Euclid had carved for him—the circles, squares, and other regular geometrical figures. For God, these most rational geometrical objects would be the perfect building materials.

The historian of mathematics Eric Temple Bell once offered the following amusing metaphor regarding Euclidean geometry and its hold on the human mind: "The cowboys have a way of trussing up a steer or pugnacious bronco which fixes the brute so that it can neither move nor think. This is the hog-tie, and it is what Euclid did to geometry."[10] The ties were finally broken in the late nineteenth century, when the Russian mathematician Nickolai Lobachevskii proved that Euclidean geometry is not the *only* geometry. In Euclidean geometry, as we were all taught at school, the angles within a triangle add up to 180 degrees. Lobachevskii showed that there are many other kinds of geometry, in some of which these angles add up to more than 180 degrees and in some less. As far as the physical world is concerned, it is a matter for experiments to decide what kind of geometry it has, and as it turns out, it is non-Euclidean, a realization Albert Einstein first made.

As I mentioned earlier, a few sociologists had hit upon the idea of non-Euclidean geometry as an effort to make sense of the apparently warped structure of social networks. In this case, however, it turned out to be the wrong idea. In contrast, small-world geometry seems to be an idea of the right sort. Here the ideas apply not to geometrical figures, however, or to the structure of space and time, but to the disorderly networks—social, economic, biological—that are the very most basic fabric of the living world.

As we are beginning to see, the small-world perspective offers a way to begin seeing order and design in apparently disordered networks of many kinds, including the human brain. Many networks that before seemed more or less random and lacking in any deep design principles are now turning out to harbor a hidden order, a disguised plan of con- siderable ingenuity.

5

THE SMALL-WORLD WEB

Mankind is a catalyzing enzyme for the transition from a carbon-based to
a silicon-based intelligence.

—Gerard Bricogne[1]

ON OCTOBER 5, 1957, disquieting news greeted readers of morning
newspapers across the United States. To the horror of military officials
and the disbelief of many American scientists, the USSR had launched
an artificial satellite into low orbit around Earth. Sputnik was now flying seven times each day over the U.S. mainland at an altitude of 560
miles. No bigger than a basketball, the satellite weighed 184 pounds,
nearly ten times as much as any device then being contemplated in the
U.S. satellite program. Whatever Sputnik was up to, it was menacing by
virtue of its existence, for it hinted that the Soviets had forged an
alarming lead in the race to space.

The real worry, of course, was not so much Sputnik itself, but rather
what it revealed: a sophisticated Soviet understanding of advanced
techniques of rocketry and missile guidance, techniques that would be
equally useful in designing and building intercontinental nuclear mis-
siles. As Republican Senator Stuart Symington suggested a few days
later, the satellite was "but more proof of growing Communist superi-
ority in the all-important missile field," and Symington went on to
sound the alarm: "The future of the United States may well be at
stake."[2] It was in this atmosphere of growing fear bordering on hysteria
that U.S. President Dwight Eisenhower crafted his response to the
Soviet surprise.

In a press conference on October 10, five days after news of the
launch, Eisenhower expressed his resolute confidence that the satellite

had not damaged U.S. military security by even "one iota." He could not say so publicly, but he knew from photos taken by U2 spy planes flying over Soviet territory that the United States was by no means losing any missile race. Nevertheless, Eisenhower's words were only soothing, and his more significant response to Sputnik came a few months later. In January of 1958, in a message to Congress, he requested funds to launch a new agency aimed at overseeing all U.S. space and weapons research. Scientists, in Eisenhower's view, were one of the United States' most valuable assets, and the Advanced Research Projects Agency (ARPA) was intended to gather their scientific might. By yoking together the often competing research arms of the various armed forces, ARPA would guarantee that the United States would never again fall behind in military technology.

In human history, nothing goes forward quite as expected. As it turned out, ARPA never did much to counter the Soviet threat in space, for within a year the creation of the National Aeronautics and Space Administration (NASA) stripped the agency of most of its funds and undermined its original purpose. One publication at the time even referred to the agency as "a dead cat hanging in the closet."[3] Forty-four years later, as a result, most people have never heard of ARPA, even though it has been in continuous existence for nearly half a century, and despite the fact that it has quietly given the world one of the most revolutionary inventions of all time.

A NETWORK OF SOME INTELLIGENCE

IN 1964, TWO years after the Cuban Missile Crisis, an American engineer named Paul Baran wrote a series of technical papers for the RAND Corporation in Santa Monica, California. Funded by the U.S. Air Force, Baran was working on the schematic design of a nationwide communications network that could withstand a considerable Soviet attack.[4] There were palpable fears at the time that the USSR might launch a surprise nuclear attack over the North Pole. The devastation of such an attack would undoubtedly be immense, and destruction of basic communication systems such as the telephone grid would severely limit the U.S. ability to respond. As a centralized system, the telephone network was highly vulnerable. Attack on just a few of its key control centers could bring the entire network crashing down.

Baran's way around this problem was what he called a "distributed" communications network, a web of computers or other communication devices linked by transmission lines, with no more than a handful of such links emerging from each. There would be no control centers of special importance. For the overall network, Baran referred to the average number of links per element as the "redundancy" of the network, and his aim was to show that "extremely survivable networks can be built using a moderately low redundancy. . . . Redundancy levels of the order of three permit withstanding extremely heavy attacks with negligible loss of communications."

Suppose a message needs to go across the network from point A to point B. The network could have a "hardwired" path or set of paths for such a message to follow, but these paths could be wiped out by the failure of a few key elements or links. In contrast, Baran's scheme would have no prearranged plan for how messages would travel. Instead, computers at each point in the network would continually make their own decisions about how best to direct a message. Each would maintain its own "routing table," reflecting how quickly or slowly recent messages had traveled over different network paths. If an enemy attack knocked out a few elements, the network computers would respond by rerouting messages around the trouble spots. The network would be adaptive—it would have a kind of intelligence all its own.

As a result, a message from point A to point B might follow any of thousands of possible paths. This would give the network great flexibility, since even a very severe attack would almost certainly leave a few workable paths in the network. As Baran wrote, "What is envisioned is a network of unmanned digital switches implementing a self-learning policy at each node so that overall traffic is effectively routed in a changing environment—without need for a central and possibly vulnerable control point."

Baran was not alone in thinking this way. By the mid-1960s, computer scientist Leonard Kleinrock of the Massachusetts Institute of Technology and physicist Donald Davis of the British National Physical Laboratory had each independently invented similar schemes, although they were motivated by other concerns. Computers in those days were bulky and expensive mainframes, and to many engineers it seemed sensible to find a way for them to talk to one another and share resources.[5] A team of ARPA-funded engineers had begun doing so, and by December of 1969, four computers had been linked together and were successfully communicating across the country. The ARPANET, as it was called, then consisted of computers at the University of California at Los Angeles, Stanford University, University of California at Santa Barbara, and University of Utah. By the end of 1972, the ARPANET had grown to include nineteen separate computers strewn across the United States.

Although no one could have known then, one of the most profound technological changes of modern times was at hand—the ironic consequence, at least in part, of Cold War machinations. The ARPANET was the original seed of today's Internet, an immense network of networks that now involves close to 100 million computers in 250 countries. The Internet, and the World Wide Web that it has spawned, are some of the most impressive creations of our civilization and together represent a milestone in social history.

As we will see, these networks are also proper objects for scientific study in their own right. As computer scientists from the Xerox Internet Ecologies Division recently wrote, "The sheer reach and structural complexity of the Web make it an ecology of knowledge, with relationships, information 'food chains' and dynamic interactions that could soon become as rich as, if not richer than, many natural ecosystems."[6] Before we turn to the task of exploring the architecture of this "Internet ecology," however, it is worth gaining some impression of just how vast and influential the Internet and World Wide Web are becoming.

WORLD CHANGES

THE INTERNET HAS doubled in size yearly for ten straight years, which amounts to an explosive thousandfold increase in the number of computers connected to it. In fact, it has grown in influence even more rapidly than did the telephone early in the twentieth century. In 1984, twenty-five years after the beginning of the ARPANET, there were more Internet hosts per capita in the United States than there were telephones twenty-five years after Alexander Graham Bell announced his invention.

Peter Drucker, professor of social science at the Claremont Graduate University in Claremont, California, is one of the great minds in business management over the past half-century. As he sees it, the computer is akin to the steam engine, and the Information Revolution is now at the point at which the Industrial Revolution was in the 1820s. Drucker points out that the most far-reaching changes of the Industrial Revolution came not directly from the steam engine itself, but as a consequence of another unprecedented invention the engine made possible—the railroad. Similarly, he suspects, it is not computers or the Internet that will be world-changing, but rather, one of their recent spin-offs: "e-Commerce is to the Information Revolution what the railroad was to the Industrial Revolution—a totally new, totally unprecedented, totally unexpected development. And like the railroad 170 years ago, e-commerce is creating a new and distinct boom, rapidly changing the economy, society, and politics."[7]

In the United States, the online auction house eBay is revolutionizing the way individuals buy and sell products. At www.ebay.com you can go online and take part in an auction for everything from used guitars or automobiles to pinball machines or houses. The company now does more than a billion dollars of business each year and, like Amazon.com, is a household name. From the *New York Times* to the *Cleveland Plain Dealer,* every major newspaper and magazine is now available online, and more than half the spending of major banks goes into developing Internet-based banking facilities. The Internet economy now supports directly more than three million workers, which makes it a larger employer than either the insurance or real estate industries. But even this is only a small part of the story. The transactions between businesses and consumers represent only one-fifth of the total Internet trade. The rest, the bulk of all e-commerce, takes place between businesses themselves.

In the words of IBM Chairman Lou Gerstner, "It's not hyperbole to say that the 'network' is quickly emerging as the largest, most dynamic, restless, sleepless marketplace of goods, services, and ideas the world has ever seen."[8] Similarly enthusiastic is Les Alberthal, former chief executive officer of Electronic Data Systems Corporation. "By now," says Alberthal, " . . . we know the revolution will never abate. In the next 10 years we will witness one of history's greatest technological transformations, in which the world's geographic markets morph into one dynamic, complex organism."[9]

You might attribute the comments of these men to their positions in leading companies that are involved intimately with the Internet. The words of corporate executives quite naturally tend to reflect the hopes of their organizations. And yet the figures back up this optimism. If current trends continue, by the year 2003, the electronic marketplace in the United States will involve more than one trillion U.S. dollars and make up nearly 10 percent of all U.S. trade. This alone will be more than the entire economic activity in the United Kingdom or Italy, and by 2006, some 40 percent of all U.S. business is expected to be run over the Internet.[10]

The Internet is altering the face of science as well. For centuries, researchers have been publishing their work in paper form in research journals, of which there are now thousands. In the summer of 2001, however, the international journal *Nature* hosted a debate on the future of scientific publishing in the light of the Internet. More than twenty thousand scientists have signed a petition to create a Public Library of Science that would make all scientific literature freely available in electronic form over the Internet. The research paper itself is changing too. No more is it merely a two-dimensional passive document; now it can be linked to videos, computer simulations, databases—you name it.

The unbridled growth of the Internet has led quite naturally to a network of staggering complexity. Both the Internet and the World Wide Web grow by the actions of millions of individuals, all with their own plans and ideas, motivated by economics, religion, or almost anything else you can imagine. There is absolutely no centrally planned structure or design, but a truly incredible tangle of computers linked to computers and Web pages connected by hypertext links. Faced with the overwhelming complexity of this network, we might well throw up our hands and despair of ever finding any simple patterns of organization within it.

And yet researchers in a variety of labs have been making maps of the Internet and the Web, and putting together wiring diagrams that capture the abstract face of cyberspace. What do these networks look like? And how do they grow? As it turns out, scientists have uncovered some surprising links to the small-world structures of Watts and Strogatz. At the same time, these networks also have some other lessons to teach.

INTERNET EXPLORERS

IN ONE OF his early RAND papers, Paul Baran considered two different kinds of distributed networks (Figure 7). One looks like the web of a fishing net, or like one of the ordered networks we saw in earlier chap-

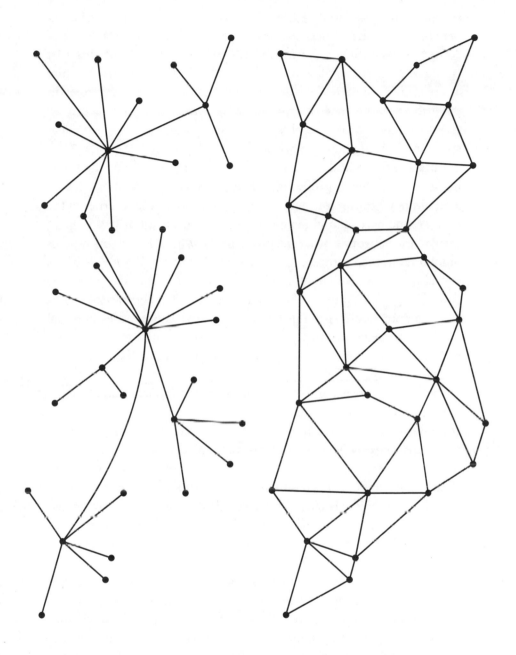

Figure 7. Diagram of two different kinds of "distributed networks," as originally envisioned by Paul Baran.

ters, while the other looks rather different. Baran referred to this latter network as a "hierarchical decentralized" network, and in a series of computer experiments, he demonstrated that networks of this sort would be more susceptible to attack than the fishnet networks. Because a few elements in such a network play especially important roles in tying the network together, an enemy targeting these "hubs" can do serious damage to the overall system. Baran clearly pointed to fishnet structures as being more "survivable."

But these theoretical ideas were from over forty years ago. The Internet is now five million times bigger than it was in 1972. What does it look like now? In view of the growing importance of the Internet to the global economy, you might guess that some central authority is keeping track of the network's structure, plans for growth, and so on. As it turns out, however, even getting a good picture of what the Internet is like is not easy. The structure of the Internet can only be discovered by painstaking exploration of the physical terrain of cyberspace, which is what computer scientists Bill Cheswick of Bell Laboratories and Hal Birch of Carnegie Mellon University have been doing for the past five years. Fortunately, they can explore the Internet from the comfort of their own offices.

Each day, Cheswick and Birch send some ten thousand small information packets out over the telephone lines to random Internet addresses, and trace the routes of each of the packets as they find their way to their targets. The technique is something like making a map of a nation's roadways by sending out an army of robots to drive over them all and then to report back on which intersections they went through. By following how packets move through the system, Cheswick and Birch can reconstruct a rough picture of the Internet's global topology. It is a topology that is gradually changing as engineers put new computers and transmission cables in place, and as the evolution of Internet traffic leads computers to route information along different channels.

In December of 1998, Cheswick and Birch produced the picture of the Internet shown in Figure 8. It reveals a snapshot of the computers that make up the Internet and the connections between them. Intriguingly, and perhaps surprisingly, the image looks more like the hierarchical network that Baran had dismissed forty years ago as being too vulnerable to attack. There are no easy answers as to why this is, for no one has been overseeing the Internet's wiring plan. The layout has evolved through innumerable accidents and reflects the decisions of

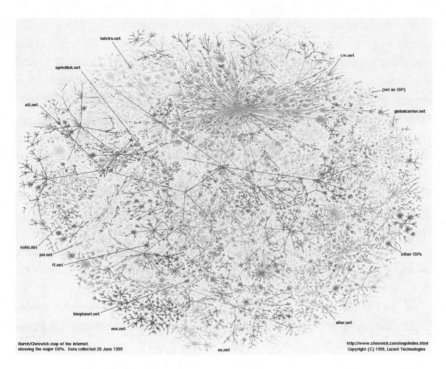

Figure 8. A map of the Internet. (Reprinted by permission of Bill Cheswick and Lucent Technologies.)

countless individuals, businesses, universities, and so on, which would lack any common theme. Nevertheless, by some mysterious principle of growth, the Internet seems to have had a certain wisdom of its own. For there is an advantage hiding in the hierarchical design that Paul Baran never knew about.

In 1999, computer scientists and brothers Michalis, Petros, and Christos Faloutsos used data for the physical network of the Internet from 1997 and 1998 to study the number of links a packet ordinarily has to traverse in going between one point and another. When a computer in San Francisco sends an e-mail to Hong Kong, or another in Helsinki tries to access information from a computer in Virginia, how many transmission lines are typically involved? As the Faloutsos team found, the answer is only about four, despite the immense size of the Internet. In fact, even if you search throughout the Internet for pairs of computers that are especially difficult to link, the number is never more than about ten.[11]

So the Internet is not only a distributed network but also a small-world network, though with an appearance that is rather different from the diagrams of Watts and Strogatz. Other researchers have repeated the Faloutsos team's studies with similar results, and also measured the degree of clustering to which Watts and Strogatz paid so much attention. For the Internet, the clustering of computers turns out to be well over a hundred times greater than would be expected for a random network.[12] So the Internet is very far from being a random network. Nor is it an ordered network as Baran envisioned. Rather, it turns out to be another small-world network that has managed to organize itself so that information can move between any two points in only a handful of steps.

Now, the Internet is not like a city where one is required to get permission from a planning board before making changes, and there is no central authority determining the layout. Anyone can link a new computer to the net, and the number of links between computers is growing rapidly, at a pace of about one each hour. So it seems surprising that the Internet has ended up having the small-world property. It is also intriguing that Cheswick and Birch's picture of the Internet is not much like the small-world networks we met last chapter, which were built by adding a few random links to an otherwise ordered network. The Internet is a small world of another sort. There is a slightly different trick at work here, and a tiny bit of mathematics can help to bring it out into the light.

CYBERSPACE

LOOK AGAIN AT the hierarchical picture of the Internet (see Figure 8). If you look closely, you will see that a few nodes have a huge number of links, far more than most, and act as hubs within the network. These hubs presumably carry an inordinate share of the information traffic. When Paul Baran was thinking about distributed networks, he did not anticipate this possibility, as he focussed on networks in which each element has about the same number of links. The Internet, by this picture at least, seems rather different.

The picture is not so easy to interpret, however, and there is a more precise way to see the difference. Suppose you look at all the nodes in the Internet, or at least a large subset thereof; count up how many have just one link, how many have two, and three, and so on; and then make a graph showing the distribution. This graph will reveal some information about the overall wiring scheme of the network. If most nodes have just a handful of links, for example, the graph will show a strong peak somewhere around three or four links. The Faloutsos team did this experiment for the network as it stood in 1998, and found a strikingly different result.

The team studied a subset of 4,389 nodes in the network, linked together by 8,256 connections, and made a graph of the sort just described (Figure 9).[13] The curve they found follows a very simple pattern that mathematicians refer to as a "power law": each time the number of links doubles, the number of nodes with that many links becomes less by about five times. This simple relationship holds right across the board from nodes with only a few links to those with several hundred, and as the Faloutsos team noted, "is unlikely to be a coincidence." The simplicity within the pattern suggests that however random and haphazard the Internet picture may look, it actually harbors a hidden order.

The same kind of order appears elsewhere as well. The Internet is not the only network underlying the Information Revolution. It is an entirely physical entity—a sprawling network of computers linked together by transmission lines. The Internet is more or less pure hardware. By contrast, the World Wide Web is rather more ethereal. This is the vast network of Web pages connected together by hypertext links those places on a Web site you can click on to be transported elsewhere. If you will, the World Wide Web is the face of the Internet, for most users interact with the Internet in this way.

Like the Internet, the growth of the Web is highly uncontrolled, almost random. Anyone can post a Web page with any number of documents and link those documents to other sites. Currently over one billion Web pages are connected together by hypertext links into one staggeringly large network. There is no obvious reason why we should expect the World Wide Web to have any structural similarities to the Internet, but it does. This network is another small-world network with a structure remarkably similar to the Internet itself.

A couple of years ago, to get a handle on the structure of the Web, physicist Albert-László Barabási and colleagues at Notre Dame University built a computer "robot" to wander over the Web and see what it could find. This was not a robot with arms or legs or even wheels;

rather, it was designed with an all-consuming passion for surfing the Web. The robot would start out on a particular Web site, go through it, and gather the names of all the links on that site. This approach is equivalent to investigating one node in the network and recognizing the links that lead from it to other nodes. The robot would then follow each of these links in turn, and repeat the procedure for each new page it arrived at.

In this way the robot could start at any particular node, and by crawling outward make a map of the Web structure. Using this robot, Barabási and his colleagues counted up how many Web pages had one link, two links, and so on. Starting with the Notre Dame site of 325,729 documents connected by 1,469,680 links, they found a pattern nearly identical to that for the Internet: the number of nodes having a certain number of links decreased by about five each time the number of links was doubled. Studying other sites such as www.whitehouse.gov and www.yahoo.com, they found the same thing.[14]

What do these simple relationships mean? Look again at the power-law curve discovered by the Faloutsos team (see Figure 9). The height of the curve at any point corresponds to how many nodes in the network have that many links. Mathematicians refer to the "tail" of the curve as that part where the number of nodes tails off toward zero. On the tail the curve reveals that few nodes have a high number of links. This is true but somewhat deceptive at the same time. For centuries scientists have been familiar with the so-called bell curve associated with "normal statistics." For example, take all the men in a large room, measure their height, and make a graph showing the distribution of their heights. You will find an average or mean height, and then a curve that drops toward zero as you move away from that average. The tail of the bell curve and the power-law curve look superficially similar, but they aren't.

Power-law curves have what are called "fat tails." That is, compared to the bell curve, the power-law curve tails off toward zero much more slowly. In the case of the Internet or World Wide Web, the fat tail implies that you are far more likely to find a node with a very high number of links than you would be if these networks followed normal statistics. You might say that these networks follow unusual statistics. In any case, there are consequences. In these networks, in fact, just a few nodes have so many links that 80 to 90 percent of the network's total number of links feed into just a small fraction of the nodes. So the power-law pattern is the mathematical face of a special architecture, an architecture that is dominated by especially well-connected hubs.

As Barabási and his colleagues concluded in the context of the Web, "The probability of finding a document with a large number of links is rather significant, the network connectivity being dominated by highly-connected web pages. . . . The probability of finding very popular addresses to which a large number of other documents point is non-negligible and an indication of the flocking sociology of the World Wide Web."

The results of these and other studies point to a common, universal architecture for both the Internet and the World Wide Web, and to a great deal of order lurking behind their apparent randomness.

Barabási's team also made an estimate of what they called the "diameter of the Web," the typical distance between documents. In

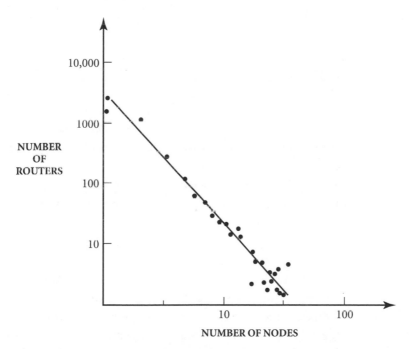

Figure 9. The distribution of Internet "nodes" according to how many links they possess. The curve follows a simple "power law" pattern. (Adapted from M. Faloutsos, P. Faloutsos, and C. Faloutsos, On power-law relationships of the Internet topology, *Comput. Commun. Rev.* 29, 251 [1999].)

other words, if you choose two documents at random, the diameter is the number of clicks you need to get from one to the other. To do this the team used their robot to build a computer model of the entire Web. The result was an estimate for the diameter of about nineteen, which points to a deep connection between the existence of hubs and the small-world architecture. Nineteen clicks may not seem terribly small, but it certainly is when compared to the billion documents in the network.

This finding is good news for the future of the Web. Based on their investigations, Barabási and his colleagues concluded that the Web diameter, D, should have a logarithmic dependence on the total number, N, of documents on the Web. In mathematical jargon, this means that even if the number N gets a lot bigger, the number of clicks needed to navigate the Web will only increase a tiny bit. "We find," the researchers concluded, "that the expected 1000% increase in the size of the Web in the next few years will change the diameter from 19 to only 21."

KINDS OF SMALL

THE STUDIES DESCRIBED in this chapter may seem a bit abstract. What do all the results mean? In the previous chapter we learned how a network can be a small world while still being highly clustered. An ordered network with a few random links thrown in is all you need. But Watts and Strogatz's recipe leaves something out: central hubs that tend to dominate network activity. The small-world network from the previous chapter shares the character originally envisioned by Paul Baran for a resilient communications network, with each element having a small number of links coming from it and none having very many more than any other. But real-world networks do not work this way.

The Internet and the World Wide Web do not quite fit Watts and Strogatz's pattern but rather achieve the same small-world end in a different way—by having a few elements with a huge number of links. In other words, there is more than one way for a network to be a small world, or, put another way, there is more to say about a network than that it is or is not a small world. If Watts and Strogatz's discovery was a first step into the world of disorderly and complex networks, then the recognition of the importance of hubs and the power-law pattern for the distribution of links is the second. And what's more, the emergence of hubs is by no means a curiosity of man-made information networks such as the Internet and the World Wide Web.

In 1999, Barabási and his colleagues Hawoong Jeong, Balint Tombor, Réka Albert, and Zoltan N. Oltavi turned their attention from the Internet and World Wide Web to the equally intricate webs of chemistry that underlie the workings of the living cell. They studied the networks of crucial biochemical reactions underpinning cellular metabolism—the basic energy-gathering and -processing functions of a cell—for forty-three different organisms. In each case, the various molecules represented the elements of the metabolic network, with molecules being linked if they participate in a chemical reaction.

Given the amount of data, the researchers used a computer to investigate the structure of the network. Sure enough, the computer revealed that these cellular networks are neither random nor ordered, and show almost precisely the same architecture as that of the Internet and the World Wide Web.[15] For every organism, the distribution of nodes according to the number of links—the number of chemical reactions in which the molecule participates—followed a power law. Cellular metabolism involves hubs as well. In the bacterium *Escherichia coli*, for example, one or two specific molecules take part in several hundred different chemical reactions involved in the bacterium's metabolism, whereas many thousands of other molecules take part in only one or two reactions. The biochemical network of cellular metabolism is also a small world, and the diameter is just about the same for all forty-three species: in every one no more than about four reactions link any two molecules.

If this kind of architecture is unusually efficient from the point of view of chemistry, then perhaps it is no surprise that the process of biological evolution by natural selection has discovered it. Millions of years of chance accidents have whittled the cellular mechanics of every organism into this structure. But then how do you explain the similar patterns discovered in the scientific research journals by physicists Sidney Redner and Mark Newman? Think of each research paper as being linked to a number of others by the references they include. Or, alternatively, think of scientists as being linked to one another by virtue of having been coauthors at some point or other. Redner and Newman unearthed power-law patterns for the distribution of links in both of these networks, again revealing a small-world architecture.[16] Studying the collaborations of physicists, biomedical researchers, and computer scientists over a five-year period, for example, Newman found that every scientist could be linked through coauthored papers to every other by only four or five steps.

As we will see in more detail in coming chapters, the same pattern holds for food webs, where species link up with one another by virtue of predator-prey relationships, and for America's most influential businessmen, who are linked by virtue of sitting together on boards of directors of major corporations. Oddly enough, the small-world architecture even turns up in the structure of human language. Last year physicists Ricard Solé and Ramon Ferrer i Cancho used the database of the British National Corpus, a 100-million-word collection of samples of written and spoken language from a wide range of sources, to study the grammatical relationships between 460,902 words in the English language. They considered two words to be "linked" if they appear next to one another in English sentences. Once again, everything tumbled out just as for the other networks.[17] A handful of words were extremely well-connected hubs, frequently appearing next to any of a huge number of other words. Words such as *a, the,* or *at* work as hubs of this sort. The typical "distance" between words in the language was less than three, just about what you would expect for a language of the same number of words that was put together at random. Nevertheless, the clustering of the network was nearly five thousand times higher than for a random network, revealing that words fall into cliques and groups, as do people within social networks. The English language is another small-world network.

Behind all of this, there are some very deep questions lurking. Not one of these networks had a designer, and yet each of them managed to arrive at much the same trick, as if they had been carefully crafted, almost for some purpose. How did these networks come to be like this?

6

AN ACCIDENTAL SCIENCE

Not one thing comes to be randomly, but all things from reason and necessity.

—Leucippus[1]

IN 1862, THE French historian Fustel de Coulanges of Strasbourg University proclaimed that "history is, or should be, a science."[2] But is history a science like physics or chemistry? Or if a science at all, is it of another sort entirely? In trying to decide, we must examine a number of issues. To begin with, it is the hallmark of physics that different researchers, working in different places at different times, eventually come to agree on the answers to important questions—from how many protons reside in an atom of oxygen to how the nuclear chemistry that fuels the Sun works. But are the methods of history adequate to ensure the same kind of convergence? Does history even work in such a way that "objective" answers to important questions exist? Even historians have expressed their doubts.

As the American historian Carl Becker pointed out in the 1930s, for example, all historians bring personal baggage to their practice of history, and this inevitably colors their interpretations of the past. Or as Becker's contemporary Ralph Gabriel put it, " 'History' is that image of the past which filters through the mind of the historian, as light through a window. Sometimes the glass is dirty; too often it is distressingly opaque. The long and sometimes unfortunate experience of mankind with history has taught the historian that the biases, prejudices, concepts, assumptions, hopes and ambitions which have contributed to the minds of his predecessors are a part of the past with which he must deal. . . . At the outset he is sadly aware that, although he may discover a few of the more obvious imperfections, his task is hopeless."[3]

The dilemma is made worse by the fact that historians have no recourse to experiment. Faced with two alternative theories, physicists can at least hope to devise some experiment to judge between the two.[4] But there is no way to reach into the past and fiddle with the details of history just to see what might happen. In writing history and in interpreting the past, each historian has to choose which facts seem most worthy of note, and this choice ultimately involves some measure of personal taste, reflecting whether he or she sees economic, political, or social forces as being most influential. No one knows the real truth, and so it is quite possible that two honest and diligent historians can arrive at distinct explanations for the same event, even though each follows "legitimate" historical methods.

Historians face another difficulty as well, one that has little to do with the subjective fallibility of humans and issues instead from the nature of historical reality itself. In the mathematical sciences such as physics and chemistry, it is possible to identify laws that appear to admit no exceptions. Einstein's famous $E = mc^2$ holds always and everywhere, for a water molecule on Mars, for the hot gases making up a distant star, or for a chunk of rock buried hundreds of miles beneath Earth's surface. The mathematical laws of quantum theory are similarly general and determine the character of the atoms that make up everything in the universe. The aim of the mathematical sciences is to discover general principles of this sort, or as the philosopher Alfred North Whitehead once expressed it, "to see the general in the particular and the eternal in the transitory."

In contrast, historians—as well as researchers in the "historical sciences" such as geology or evolutionary biology—have a far harder task identifying laws that would hold without fail. Too many accidents and chance events force their way onto the stage, each leaving a mark on the unfolding future, and so explanations take the form not of references to general laws, but of stories that connect events together and tell how things came to be. An explanation might relate how event A led to event B, and how B led to C. As a consequence, it becomes clear that had A never taken place, neither would have B or C. Had the German army not invaded Russia in the summer of 1941, for example, the Battle of Normandy might never have taken place—and certainly not as early as the summer of 1944, for many of the German divisions diverted to Russia would have remained in France, presenting the Allies with a truly formidable Atlantic wall.

The evolutionary biologist Stephen Jay Gould argued, quite rightly, that "contingency" of this kind lies at the very core of history. "I am not speaking of randomness," Gould wrote, " . . . but of the central principle of all history—contingency. A historical explanation does not rest on direct deductions from the laws of nature, but on an unpredictable sequence of antecedent states, where any major change in any step of the sequence would have altered the final result. This final result is therefore dependent, or contingent, upon everything that came before—the unerasable and determining signature of history."[5] But if contingency is king of history, and if all historical "science" must therefore rely on narrative storytelling as a method of explanation, this puts us in a rather strange situation. Questions concerning the evolution of social networks, the Internet, the molecular innards of the living cell, and the structure of human languages certainly fall within the domain of history. There are detailed stories to be told about each of these, and we would expect the stories naturally to have little in common. Each of these networks is the product of a wholly unique history, and the forces that have influenced and crafted the structure of the cell, for example, would appear to have absolutely no overlap with the technological and economic forces that have built another such as the Internet.

Nevertheless, as we have seen, each of these networks reveals the very same lawlike architectural design. Each is a small world while also being highly clustered. Furthermore, unlike networks conforming to Watts and Strogatz's original recipe, each is dominated by hubs—a few extremely well-connected individuals, Web sites, or what have you. What's more, we are not just talking about some loose, qualitative similarity. This feature has a specific mathematical signature: the power-law or fat-tail pattern for the distribution of elements according to how many links they have. And this signature turns out to be nearly identical from one kind of network to the next.

What we see then is a kind of natural order that for mysterious reasons seems to well up in networks of all kinds and that does so despite the complexities of their individual histories. How can this be? Gould is certainly right that contingency is "the unerasable and determining signature of history," but this is not to say that there is nothing to history but contingency. In biology, Charles Darwin's idea of evolution by natural selection offers an extremely powerful organizing framework within which historical accidents take place. In the context of networks, there must also be some deeper principle at work.

In chapter 7, we will see what this principle is. But first we will explore more widely how it is possible that pattern and order can emerge from nowhere, sometimes even from chaos, and regularity can tumble out as the consequence of a long string of mere accidents. As we will see, there is a flip side to contingency, and more form lurking within history than we might naively suspect.

PANNING FOR PATTERNS

IF YOU HEAT water gently in a pan, it will begin roiling and flowing about, as the heat streaming up from below stirs the liquid into motion. To most of us this is hardly worthy of note; it takes place in the kitchen every day. But in 1901, a twenty-seven-year-old Frenchman named Henri Bénard recognized that the liquid will not move if the temperature of the burner is low enough, and from this he drew a striking conclusion. If the liquid moves when the temperature is high, and doesn't when it is low, there must be some crucial temperature in between at which the movement first begins. As a student at the College de France in Paris, Bénard set out to discover how this "beginning" takes place.

Bénard set up an apparatus that would heat the pan uniformly, without creating any hot or cold spots, and rigged up a camera to take pictures from above. In order to see the liquid moving more easily, he also added a tiny amount of dust to it. Bènard then set to work. At first, with the heat turned very low, the liquid remained at rest within the pan, just as expected. Bénard slowly increased the heat, but still nothing happened. Bit by bit he turned it up further, waiting for something to change, and finally it did: the liquid suddenly and abruptly burst into motion, forming a spectacular and almost perfect arrangement of hexagons (Figure 10). Turning the heat down, Bénard could make the pattern go away; increasing it again, he could bring it back.

Over the next few weeks, Bénard studied the liquid more closely and soon learned that within each hexagon, warm liquid was flowing upward in the dark center and cooler liquid was sinking at the boundaries. Somehow, the water was managing to organize itself into this striking pattern, as if each part knew what the others were doing and responded accordingly. To his disappointment, Bénard never managed

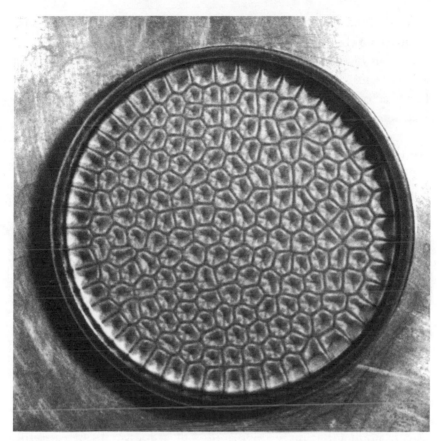

Figure 10. A hexagonal pattern of fluid flow in a shallow tray of liquid that is heated from below. (Image courtesy of Manuel Velarde, reprinted by permission.)

to explain how and why this happened, but sixteen years later, the British physicist Lord Rayleigh thought he could.

Near the bottom of the container, where the heat is applied, the warmer liquid should expand and become less dense than the cooler liquid above. In principle, it should then rise like a hot-air balloon, while the heavier, cooler liquid should sink. Rayleigh pointed out that this would indeed be the case, except for the friction within the liquid —known as *viscosity*—which acts to retard the liquid's motion. Because of viscosity, warm liquid will not rise and cold liquid will not sink unless the heating is sufficiently powerful, just as Bénard had observed.

In a theoretical tour de force, Rayleigh used the mathematical equations of fluid physics to back up his argument, and even demonstrated that the natural way for the liquid to break out into motion would be by adopting the perfect hexagonal pattern. His theory was so natural and impressive that scientists accepted it for half a century, which is ironic because it was incorrect. As it turns out, Rayleigh in his calculations had considered a closed container that was filled to the very top. His water had no exposed surface. Bénard, in contrast, had experimented with water in an open pan, with its surface exposed to the air. For technical reasons, this means that Rayleigh's explanation does not really apply to what Bénard observed.[6]

Nevertheless, the general spirit of Rayleigh's thinking was spot on, for he recognized that a crucial battle was at work behind the scenes of the experiment, and that this battle was at the root of the sudden emergence of order where before there was none. In the heated liquid, there is a competition between the forces of viscosity, which tend to keep the liquid stationary, and the heat, which tends to drive it into motion. When viscosity wins out, the liquid remains perfectly still and featureless. But when the heat takes over, the featureless perfection dissolves, and something emerges where once there was nothing.

Bénard's experiment offers a striking illustration, perhaps the simplest possible, of the spontaneous emergence of order within a network of interacting elements, in this example, molecules of water. You would naturally anticipate that uniform forcing of a uniform system would lead to nothing but more uniformity, but it doesn't. To be sure, this does not teach us anything directly about the complex and disorderly networks we have been exploring. As we will see, however, patterns in a pan are by no means a strange and isolated curiosity.

ACTS OF CREATION

IN 1831, LONG before Bénard did his experiment, the British physicist Michael Faraday was experimenting with a container of water, gently shaking it up and down, when he discovered a similar surprise. When the shaking was sufficiently weak, nothing much happened—the liquid remained in a flat and featureless layer. But more vigorous vibration destroyed the uniformity, and the liquid suddenly organized itself into a series of ridges and troughs forming stripes or checkerboard patterns.

The abrupt dissolution of uniformity into a pattern is somewhat akin to the mythical Act of Creation, for in the process, something

Figure 11. Various patterns produced by the vertical vibration of a thin layer of sand. (Image courtesy of Harry Swinnery, Paul Umbanhowar, and Daniel Goldman, reprinted by permission.)

arises from nothing in a way that is almost more basic than physics itself. Put ordinary sand in an empty box and shake it, and you can see the same thing. When the shaking becomes violent enough, the originally flat layer of sand suddenly piles up, sometimes forming a striped pattern of ridges separated by valleys and at other times forming ridges that link up to form beautiful networks of squares or hexagons. The sand manages to make the pattern all by itself (Figure 11).

In each of these cases, the fundamental battle that Rayleigh identified is at work: some forces try to create a pattern while others try to wash it away. And the battle is by no means confined to the carefully controlled setting of the laboratory. The picture in Figure 12 shows rocks in Svalbard in the Norwegian Arctic which look as if they have been pushed up into circular piles by the painstaking efforts of humans. But this spectacular organization has welled up on its own. The bare soil in the center of each circle is about 2 to 3 meters across, and the circular pile about 20 centimeters high. The rocks arrive at this pattern as the result of the repeated freezing and thawing of the ground over thousands of years.

Figure 12. In the tundra, cycles of freezing and thawing work stones into circular patterns. (Photos taken in Western Spitsenberg by Mark Kessler, reprinted by permission.)

All this information is intriguing, but these regular geometrical patterns would appear to have very little to do with the complex and messy structures of the Internet or the living cell. There are no hexagons in the cell, and the links between computers in the Internet do not form a nice, regular framework. The reason is history—and its signature, contingency. In the laboratory, physicists can repeat Bénard's experiment one thousand times and always find the same hexagonal pattern; the same goes for the shaken containers of water or sand. Here we have order emerging on its own, but in a way that pays no attention to historical accidents and is fully in keeping with the timeless equations of physics. Loosely speaking, under the appropriate conditions, the water or sand "prefers" to adopt the patterned arrangement and will do so every time. Reach in with a spoon and stir the water in Bénard's pan, or disturb the sand with your finger, and in a short while the same pattern will re-emerge. The buffeting forces of history leave no trace on the future.

In absolute contrast, countless unorchestrated historical events have left their traces all over our social and ecological networks, the World Wide Web, and so on. When Amazon.com started selling books over the Web, they not only launched a site that soon developed into a hub to which many hundreds of thousands of other sites have links, but also set loose an idea that has stimulated many others to start online book-selling services. Had Amazon.com never existed, the Web would be dramatically altered in many of its details. The biochemistry of the bacterium *Escherichia coli* has similarly been affected by a long string of genetic mutations, every one accidental and all of which have left their distinct traces on the structure of its present biochemistry. In the evolution of networks, history is tremendously important.

Even so, all the studies discussed in the previous chapter point to some kind of unifying organization in these networks, an order that is not so readily apparent as Bénard's hexagons and that does not "look" like order as we usually think of it. But as we know, order and meaning often depend not only on the nature of physical reality but also on the eye and mind of the beholder. A conversation in Hungarian can seem perfectly sensible and ordered to one person, and sound like nonsense to another. The hexagonal patterns that welled up in Bénard's pan seem obviously ordered to everyone, but in addition to these kinds of patterns, thought to be so perfect by Greek thinkers such as Pythagoras and Plato, there is another, more subtle kind of order. To perceive this kind of order, and to comprehend its origins, we need to look in another direction.

Bénard's experiment illustrates how order and pattern can emerge from featureless nothingness, from uniformity. What about order from pure chaos and randomness? This is possible too, proving that striking order can emerge even in the face of history and its contingencies.

ORDER IN DISGUISE

THE WATERS OF the great Mississippi River start out as rainfall in thirty-one states stretching from Wyoming in the West to New York in the East. Starting in Minnesota, the Upper Mississippi flows southward and near St. Louis absorbs the vast waters of the Missouri River. Further east, in Pennsylvania, the Allegheny and the Monongahela Rivers join at Pittsburgh to become the Ohio, which then flows westward to join the

Mississippi near the southern tip of Illinois. Ultimately, the Mississippi drains more than 3 million square kilometers of U.S. territory and carries this entire load southward to New Orleans and the Gulf of Mexico.

Scarcely anything can look less planned and lacking in design than the drainage basin of the Mississippi, or of any other great river network for that matter, either great or small, from the Amazon in South America and the Congo in Africa to the less famous Fella River in northern Italy (Figure 13). Traveling upstream along any river, tributaries branch off to one side or the other as if positioned by some great natural lottery. This should not be a surprise. The particular layout of any river network reflects the specific geophysical history and character of the region: the climatic patterns of rainfall, the kinds of rocks and minerals that make up the mountains and plains, and so on. Each network is a fingerprint that captures all the intricate details of the rich life history of its unique place on Earth.

Nevertheless, this random, haphazard appearance disguises a hidden order. If every river network is unique, they are also in many respects

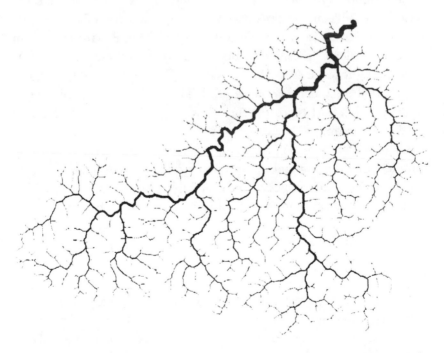

Figure 13. The structure of the Fella River network in northern Italy. (Reprinted by permission from Ignacio Rodríguez-Iturbe and Andrea Rinaldo, *Fractal River Basins* [Cambridge University Press, Cambridge, 1997].)

deeply similar, indeed, even identical. We only need the right perspective to see how.

Starting with any particular stream segment, you might decide to
work backward, moving upstream, tracing all the tributaries that lead
into it. By doing so, you would come to see all the branches of the great
tree of tributaries that ultimately feed into this one stream, and could
estimate the total land area that this particular stream drains. Scientists
who study rivers call this the *catchment area* of the stream, and clearly,
segments upstream will have relatively smaller catchment areas than
segments downstream. In the case of the Mississippi River, for example,
the final stretch that pours into the Gulf of Mexico has an enormous
catchment area—after all, it drains about 41 percent of the entire continental United States.

For a great river system such as the Mississippi, there may be tens of
thousands of individual stream segments in the entire network. But just
suppose that by using aerial photographs and radar images taken from
satellites, you could analyze the entire network in the following way:
For every stream segment in the network, you calculate its catchment
area, the entire area upstream that it drains. Then you determine how
many streams drain 10 square kilometers, how many drain 20 square
kilometers, and so on, continuing up through 1,000 square kilometers
and going right up to the one final river segment through which all the
waters finally flow. Then make a graph to display your results—number
of segments versus the area they drain.

In principle, your results might show that there is a "typical"
drainage area for a segment in the network; that is, the curve might be
sharply peaked about an area that would represent the "usual" drainage
burden for a section of river. This is what you would find, for example,
if you weighed several hundred apples from a tree, a curve that would
be peaked about the typical weight. For river networks, however, many
scientists have completed the investigation described and discovered
something very different: the distribution does not have any peaks
whatsoever but instead follows a pattern we introduced in the preceding chapter: a power law. As the area drained by a stream doubles, the
number of such streams falls by a factor of about 2.7. If there are 100
streams draining 1,000 square kilometers, then there will be only about
37 draining 2,000 square kilometers, and so on.

To begin with, this finding represents an impressive and rather unexpected simplicity for a network that seems to have little if any order. But

there is more. If this pattern held true for just one river network, you might put it down to some quirky feature of the local geology. But scientists over the past few decades have uncovered this pattern in the Nile, the Amazon, the Mississippi, and the Volga—indeed, in every river network they have ever studied. It is not the exceptional result of some peculiar local geology, but points to some deep organizing tendency behind the apparent disorder of all these networks. If a river network really were "just" the product of a long string of historical accidents, with waters eroding channels here and there without regard for any overall plan, then this power-law pattern would not exist.

In coming to an understanding of why the Missisippi River basin looks exactly as it does, and not otherwise, we would need to take into account everything from weather patterns and climate change over tens of thousands of years to the geological details of the terrain in which the river has emerged. Earthquakes have altered the path of the river on numerous occasions, and these too would need to be mentioned; quakes have created new lakes and branches of the river, and even made it flow backward for a few days following the great New Madrid quakes of 1811–1812. But the power-law pattern suggests that there is a deeper reality and something else to discover behind these details. If the same pattern turns up in the Mississippi, the Nile, and the Amazon, details of geology and weather must have little to do with it. Some universal process must be at work, a process that lives behind the details and makes all river networks turn out alike.

A PROCESS BEHIND HISTORY

THE RUSSIAN POLITICIAN and revolutionary Leon Trotsky once suggested that the laws of history are difficult to perceive because they are so effectively obscured by the accidents of history: "The entire historical process is a refraction of historical law through the accidental. In the language of biology, we might say that the historical law is realized through the natural selection of accidents."[7] In this view, despite the myriad and contradictory accidents that push history this way and that, there stands behind the entire confusion a meaningful pattern and progression, a deeper historical process that is constant in its action. Trotsky argued that uncovering and describing this process is what a science of history ought to be about. It is unclear whether this view is really profitable when applied to human history. As we will see, however, it certainly makes a wealth of sense in the context of river networks, and just possibly for many other kinds of networks as well.

The way to trace the outlines of the process behind river network formation is to work backward, first by ignoring almost all the factors that might conceivably affect the evolution of a river network and starting with a few of the most obviously important. You can see if these are enough to explain the pattern; if not, then the network formation is more complicated and other factors need to be included. Over the past decade or so, physicists Andrea Rinaldo of the University of Padua in Italy and Ignacio Rodríguez-Iturbe of Texas A&M University have followed this strategy and have discovered that the complex patterns of the world's river networks are far simpler than anyone would have guessed.[8]

To model river network formation, Rinaldo and Rodríguez-Iturbe began with a simple picture: When rain falls, it always runs downhill. The layout of the land guides the water's movement. Over the years the landscape changes, for as water flows over the ground and through the channels already formed, it causes erosion. Wherever the land is steeper, the water flows faster and causes more erosion, digging out channels that then can carry even more water. Using a computer, Rinaldo and Rodríguez-Iturbe started with a model of a random landscape having no particular pattern whatsoever, and let water fall on it uniformly. Then they kept track of the erosion and how this modified the landscape. In this way, they could re-create the evolution of a river network in only a few minutes, rather than millions of years. Indeed, they could easily re-run the tape of history many times, generating hundreds or thousands of river networks. What is surprising, even astonishing, is how accurately the results reflect the character of real river networks, despite the extreme crudeness of the model itself.

In their computer experiments Rinaldo and Rodríguez-Iturbe did not include any details of the physical properties of the rocks or soil, details that would influence the speed of erosion. They also did not worry about the fact that rainfall patterns are not usually uniform, and they left most other aspects of reality out as well. Miraculously, none of these factors seemed to matter. The networks that emerged from the computer (Figure 14) matched up well mathematically with the patterns of real river networks. Consider the distribution of streams according to the land area they drain, for example. In the model, each

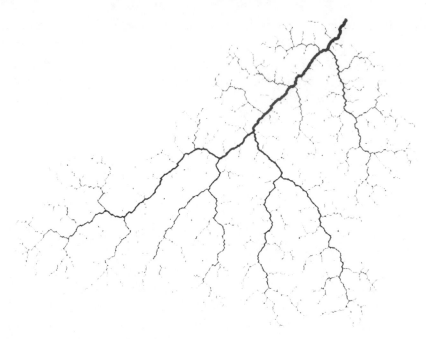

Figure 14. River network generated by a simple erosion process. (Reprinted by permission from Ignacio Rodríguez-Iturbe and Andrea Rinaldo, *Fractal River Basins* [Cambridge University Press, Cambridge, 1997].)

time the size of a basin doubled, the number of such basins fell by a factor of about 2.7, which is exactly the same as in real networks. Despite the outrageous crudity of the process at the core of their model, it fits the facts—the truth behind river network formation is far simpler than it seems.

This remarkable correspondence between computer model and reality brings us to the real point of the power law, the very same pattern we found in the context of the Internet and several other networks. In each run, the computer generated a distinct river network. In 100 consecutive runs, the networks were never quite the same in terms of the detailed layout. Accidents and historical contingency play an important role here. Nevertheless, the resulting networks always follow the power-law pattern, which points to a universal feature of river network formation.

The power law implies that if you magnify any small portion of a river network, you will get a pattern that looks much like the whole. In other words, the network is not nearly as complex as it appears. Innumerable accidents may make every river network unique, and yet what goes on at one scale is in every case intimately connected to what goes on at another. This feature, which reveals a hidden simplicity in the structure of all river networks, is known as *self-similarity*, and structures of this sort are sometimes called *fractals*. The real importance of the power law is that it reveals how, even in a historical process influenced by random chance, lawlike patterns can still emerge. In terms of their self-similar nature, all river networks are alike. History and chance are fully compatible with the existence of lawlike order and pattern.

So there is more to the historical sciences than storytelling. In explaining the existence and positioning of one particular branch in a river network, perhaps we have no choice but to wade into all the historical accidents that ultimately led to its evolution. That branch certainly could have been elsewhere and might trace its origin back to a big storm one night long ago that washed out a brand new channel. If history were run over again, the storm and its water might have gone elsewhere and the entire river network in its details would be different. Nevertheless, the network as a whole would still have the very same fractal character and would satisfy the same power law that reflects its globally organized self-similar architecture. This pattern shows up inevitably and reveals, in Whitehead's phrase, "the general in the particular and the eternal in the transitory."

NATURAL NETWORKS

THERE IS AN even simpler illustration of the way patterns can emerge out of a stream of historical accidents. Physicists in recent years have developed a class of mathematical games for what they call *diffusion-limited aggregation* (DLA), an ugly term for a beautiful process that generates spectacular patterns. More clearly than any other process, perhaps, this mathematical game illustrates the meaning of a power law and points out how to uncover organization in a complex and disorderly world.

The process is as follows: You start with a single molecule sitting alone by itself. Then you let another molecule wander in from afar, approaching from a random direction, and following a random path. If this molecule runs into the one already there, it sticks. If it misses and wanders off, then it doesn't. The idea is to repeat the process millions of times and see what happens. Each approaching molecule either hits and sticks to the cluster of molecules already there, or doesn't hit and doesn't stick. You can imagine watching the game running on a video screen.

You might expect the cluster to grow into an uninteresting blob. Instead, the structures that form are exceedingly complex (Figure 15). Once the structure has developed its branched arms, they act as screening obstacles and tend to grab incoming molecules. So the next molecule almost always gets attached near the end of one of the longer arms, making it still longer. Rarely do the molecules find their way deeper inside the cluster. The growth ends up following a process in which the rich get richer, or the longer arms get longer even faster than the shorter ones. The resulting structure is a fractal, and it has a degree of self-similarity—as in any river basin, the pieces look very much like the

Figure 15. A cluster formed by the process of "diffusion-limited aggregation." (Image courtesy of Paul Meakin, reprinted by permission.)

whole. And this self-similarity is reflected in a power law. There are branches in the cluster of all sizes, some tiny, others large. If you count how many there are of all the various sizes, you would find that every time you decrease the size of the branch by a factor of two, the number of such branches would increase by a factor of about three.

Just like river drainage networks, these DLA clusters do not appear to have any organization at all. Like the networks we have been exploring in earlier chapters, they grow through countless accidents, each of which leaves a permanent trace. In this case, the accidents are truly random, by definition. But a kind of order is still lurking. Run the experiment ten million times and you will find ten million unique structures. Nevertheless, the clusters will always "look" roughly the same; that is, your eye and brain will pick out some deep similarities, even though the details are always different.

So if contingency is "the signature of history," the power law is the equally distinct signature of a deeper kind of order that can well up even in the face of history's disruptive accidents. With this understanding, it is time to return to our networks, such as the Internet and the living cell, to see if this perspective can be of any use in explaining their architectural similarities. As we have seen, scientists have discovered power-law patterns in these networks, suggesting that despite all the manifold differences in what makes these networks grow and evolve, they may share some common core process of growth, a process that affects our world at many levels and in many different settings, even though its workings have hitherto been unsuspected.

7

THE RICH GET RICHER

The researches of many commentators have already thrown much darkness on this subject, and it is probable that, if they continue, we shall soon
know nothing about it at all.

—*Mark Twain*[1]

Whats accounts for an outbreak of riot? At 8:30 P.M. on the
evening of Sunday, April 15, 2001, local residents in Bradford, England,
reported a large fight in a pub called the Coach House. West Yorkshire
police immediately sent 130 officers in riot gear to the pub, where they
were confronted by youths throwing bricks and petrol bombs. The pub
was ablaze and its windows smashed, cars in the roadway were burning,
and rioters were looting shops nearby. Apparently, the riot erupted after
two people started a fight in the pub. This incident touched off more
rioting on the following night, and again each night for the next week,
as gangs of youths roamed the city, pulling motorists from their cars,
beating people in restaurants and pubs, and burning and looting as
they went. The unrest persisted for the next three months as right-wing
activists from across England flocked to Bradford to stir up more trouble. In one particularly violent evening in June, local police needed the
aid of eight neighboring departments when one thousand rioters
nearly overwhelmed five hundred officers.

Could any of this violence have been anticipated? As West Yorkshire
Police Superintendent Mark Whyman commented after the first night
of rioting, "There are tensions in Bradford. It is a multicultural society.
But our work hasn't given us any indications of the problems that we
saw last night."[2] Racial antagonism was certainly one of the contributing factors, as were difficult economic conditions in Bradford. These

general influences go a long way toward explaining why riots were possible, perhaps even likely in the spring of 2001. But why did that one particular fight on April 15 light the fuse?

When faced with the mysteries of crowd behavior, people often talk about the madness or irrationality of crowds, about herding behavior and mass psychology—and it is true that the actions of any crowd are immensely difficult to predict. But at least part of the reason behind the capriciousness of crowds is not actually so mystifying. In the late 1970s, with a little mathematics, Mark Granovetter made the point in a particularly illuminating way.

Granovetter started with the idea that we all have a "threshold" for joining a riot. Most of us would not start a riot over nothing, but we might join in under the right circumstance—if we were, in some sense, "pushed" far enough. Of one hundred people standing around in a pub, one person might join a riot if ten others were already smashing things up, whereas another person might require sixty or seventy people to be rioting before joining the crowd. The level of someone's threshold would depend on their personality, and on how seriously they take threats of punishment, for example. Some might not riot under any conditions, no matter how many others joined in, while a rare few might be happy to trigger the riot all on their own.

Obviously, a person's threshold would be rather difficult to determine in practice, but this is not so important. From a conceptual point of view, we all must have some threshold, this being the point, as Granovetter put it, "where the perceived benefits to an individual of doing the thing in question (here, joining the riot) exceed the perceived costs." And what is intriguing is how this threshold—or rather, the fact that thresholds vary from person to person—affects the complexity and unpredictability of a group's behavior.

To illustrate, imagine that the thresholds of the one hundred people in the pub range from 0 to 99, with each person's threshold being unique. One person has threshold 0, another 1, another 2, and so on. In this case, a large riot will be inevitable. The "radical" with threshold 0 will kick it off, then will be joined by the person with threshold 1, and the riot will grow like wildfire, eventually sucking in even the people with very high thresholds. But notice how delicately the outcome depends on the character of even a single person in this chain. If the person with a threshold equal to 1 had a threshold of 2 instead, then after the first person started smashing things up, the rest might simply stand by watching and perhaps even call the police. With no one willing to be the second person to riot, there would be no chain reaction.

So a tiny difference in the character of just one person can have a dramatic effect on the overall group. As Granovetter noted, however, a newspaper reporting on the two different cases would probably miss this subtlety. In the former case, the story might say that "a crowd of radicals engaged in riotous behavior," while in the latter it might report instead that "a demented troublemaker broke a window while a group of solid citizens looked on."[3]

Granovetter's toy model is not meant to explain how riots start, and it will not help officials relieve the tensions that make riots more likely. But it does shed light on why the behavior of a group is so hard to predict. For what the group does as a whole reflects not just its average makeup, but the precise details of how the various thresholds of all its members link together. Indeed, we all have probably seen the same effect in other settings. Think of a group of students deciding whether to do their homework or to go out for a few beers. Even a devoted student might be swayed if his five friends decide to go out. Or think of some friends at a party deciding when it is time to leave. We all have seen a party suddenly disintegrate after one person finally decides it is time to go.

The point is that we can learn a lot by starting simple and attributing to people quite rudimentary patterns of behavior. As Albert Einstein once said, the whole point of scientific thinking is to make things "as simple as possible, but not simpler." And this turns out to be true as well in trying to make sense of how the World Wide Web, the Internet, and the other complex networks we have been exploring have earned their peculiar architectures, and especially why those architectures turn out to be so similar.

LAWS OF SURFING

SUPPOSE YOU RUN a small furniture-making business in a tiny village in the middle of England. You make excellent handcrafted furniture from reclaimed pine, and with your limited financial resources, you want to market your products. The Web seems like an excellent and economical way to do that. So, you design your own Web page, and you are about to put it online. You have just one final detail to worry about: to which other Web sites should you link your own? Links add character and depth to a site and represent your own local road map of the Web, designed to help your visitors locate useful and interesting information. How do you go about making this map?

To go with their new bed, some of your clients might want to purchase a mattress from a well-known mattress manufacturer in a neighboring village. For their convenience, you might provide a link to the mattress manufacturer's Web site. You also know that some people who buy your furniture do so because you use reclaimed pine, recovered from old buildings and such, rather than virgin pine from new trees. So, you might want to provide links to popular ecologically minded sites where they can learn more about reclaimed pine and how you are helping the environment. And just for fun, to give your site a personal feel, you might put a few links to sites that you just happen to like.

Your links will reflect your particular business, where you live, and lots of other personal details. One thing, however, is certain: you will not add links to sites you have never heard of. You might learn about sites from friends, by reading magazines or newspapers, by listening to the radio, or by surfing the Web yourself. In any case, you will tend to hear about the more popular sites rather than the less popular ones. This insight, in itself, is not terribly earthshaking. But it does suggest something about the way the network of the Web should grow: the more popular sites, those with a large number of links already pointing to them, should tend to grow the fastest. The more links a site has pointing to it now, the more it should gather in the future. The rich should get richer.

But wait a minute. Approximately 85 percent of all people use major search engines such as Yahoo, Infoseek, Altavista, and Google to find what they are looking for. By searching the Web this way, can't you find pages regardless of how popular they might be? Doesn't this spoil the pattern? In fact, no. On the afternoon of June 16, 2001, I used Google to search for pages that might have information related to the phrase "Internet architecture," and the engine came back in a snap with links to a stupefying 1,660,000 pages. This number sounds rather thorough and impressive. Most of these pages, however, were largely irrelevant. The depth of the search is illusory for another reason as well.

In 1999, Steve Lawrence and Lee Giles of the NEC Research Institute in Princeton studied the search engines and found that none at the time were covering more than 16 percent of the Web.[4] The Web is growing at a terrific pace—with expectations for another tenfold increase in size over the next few years—and the search engines just cannot keep up. It is as if the mapmakers of the Middle Ages were trying to map a world that was expanding like a balloon, creating new territory so fast that no amount of exploration could keep up. For searching the Web, you can do a little better by using Metacrawler, a resource that combines the results of the main search engines. But even Metacrawler was then only covering about 50 percent of the Web.

Lawrence and Giles also found, somewhat disturbingly, that new pages posted on the Web often do not show up in the search engines for several months. The reason? Again, it is the influence of the popular. Most of the search engines "index" Web pages based on their popularity. When you search with Google, for example, you do not really search the Web but Google's index of the Web. This index is updated frequently in an attempt to keep pace with Web growth, but there is a bias: the more popular the page, the earlier it gets indexed. Therefore, new pages, even those with excellent content, face an uphill struggle in becoming known. So whether you use a search engine or not, if you start a Web site and add links to other sites, they will tend to be popular sites.

Hence, the rich get richer, and the more popular grow more popular still. So what? Does this say anything worthwhile about how the network of the World Wide Web grows? As it turns out, it does. Indeed, it points to what may be one of the most basic and important principles of network architecture in all nature.

THE ART OF ATTACHMENT

THERE IS NO history in the small-world recipe that Watts and Strogatz discovered. If you start out with an ordered network of ten thousand elements and add in a handful of long-distance links, you end up with a network still having about ten thousand elements. The small-world networks of Watts and Strogatz do not start with just a few elements and then grow over time, gaining complexity and one day reaching maturity in the small-world architecture. From the mathematical point of view, this lack of growth is no problem. But the Internet and the World Wide Web are real-world networks. They started out small and grew into the massive networks they are today. And the way they are today must reflect how they have grown.

Although the basic idea of Watts and Strogatz spearheaded the scientific invasion of the world of complex networks, it does not really explain how these networks have come to be. Could the rich getting richer—a mechanism that we might call "preferential attachment"—have something to do with it? At the outset, the idea seems tempting. In the acting world, after all, new and unknown actors tend to start their careers in supporting roles alongside more famous actors. Consequently, in the acting network, a new actor will tend to be linked more frequently to well-known actors rather than other upstarts. Likewise, when scientists write a new paper, they are more likely to cite well-known papers in their field, those that already have been cited many times before, rather than obscure papers of which few have heard. Again, the rich get richer.

In 1999, physicists Albert-László Barabási and Réka Albert tried to find out where this basic mechanism of the rich getting richer would lead. That is, if a network starts out small and grows by preferential attachment in the simplest way possible, what comes out? At the University of Notre Dame, Barabási and Albert invented a simple scheme to find out.

Imagine a network in its early stages, when it has only a handful of elements—these could be Web pages, actors, scientific papers linked together by citations, or what have you. To make the network really simple, we can begin with only four elements. Now, suppose the network grows by taking on a new element every so often, and that new element connects up randomly to a couple of the elements already there. To picture the process, imagine four rocks in a grassy field. Each day we carry a new rock to the field, drop it on the ground, and attach it with strings to two of the other rocks there, choosing our targets at random. So far, this scheme simply adds new elements to the network, and wires them in place haphazardly.

But now let's add a slight bias to the linking part of the recipe. Suppose we have been carrying rocks and attaching them for many days. One day we arrive at the field with a new rock, and we look around at the others. We notice that many have two or three strings attached, some have seven or eight, and a few have ten, even fifteen. In choosing a pair of rocks to tie the new one to, suppose we don't work totally at random but give an advantage to the rocks that already have a large number of strings attached, so a rock having six strings, for example, will be twice as likely to get another string as a rock having just three strings, and so on.

If we think of these elements not as rocks but as Web pages, then we capture the idea that popular pages are better known than unpopular ones and therefore have an advantage in gaining more links. Or, if we are talking about research papers, this means that a new paper coming out will tend to cite well-known papers rather than obscure ones, and so on. Of course, the scheme is so dead simple that it might well produce little of interest. And certainly, in the short run, that is so. But when history has a chance to act, something more striking emerges.

Repeating the procedure a million times, we will have added a million elements to the original four, to give a total of a million and four. At each step we also will have added two new links and will now have more than two million links in the entire network. By all accounts, the result is a visual mess (Figure 16). If we do the experiment again, since the links are always chosen at random, the new network will be another mess, different from the first in all its details. If we run the experiment a thousand times, we will get a different mess each time.

Barabási and Albert ran the growth process many times in the computer, generating a huge number of networks. Sometimes they changed the number of elements they started with, beginning with thirty-seven or twenty-six rather than four, or adding in seven or twelve links at each step rather than just two. Remarkably, however, none of the variations seemed to matter in the long run. The networks that grew were always the same in terms of their basic architecture: every one was a small

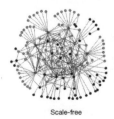

Scale-free

Figure 16. A "scale-free" or "aristocratic" network. (Image courtesy of Albert-László Barabási, reprinted by permission.)

world, with only a few steps needed to go from one element to another. The networks also were highly clustered and revealed the "hub" feature of real-world networks. As icing on the cake, Barabási and Albert studied the distribution of elements according to their number of links and found the telltale power-law pattern: each time they doubled the number of links, the number of elements having that many fell off by a factor of about eight.[5]

So the rich-get-richer mechanism has more in it than you might think. It turns out to be a natural engine of small-world architecture and may well lie behind the structures of many of the other networks we explored in earlier chapters. As with river networks, there appears to be a great simplicity behind the apparent complexity. The key is little more than pure relentless mathematics, perhaps aided by the psychology that draws us toward the already popular and the well connected.

GROUPTHINK

THIS PATTERN OF the rich getting richer is well known in every industry where names and labels sell things. One very effective way to sell more product is to become popular in the first place, then it becomes easy. Very few people ask for a cola; they ask for a Coke. If a hundred competing brands showed up on the market tomorrow, most people would still buy Coke without even trying the others. The rich get richer, in part, because people find it easy to fall back on name recognition in making their choices, opting for the film starring Michael Douglas over another starring Jason Paulick. And if you are making a new Hollywood blockbuster, with millions of dollars invested before filming even starts, who do you want to play the lead roles? It does not take too much imagination to realize that such choices do not come down to acting skills and suitability for the part alone. Advertising potential based on name recognition is crucial. Put Julia Roberts, Leonardo DiCaprio, Sean Connery, or Kevin Costner in your film, and you are immediately going to draw more people in, even if the film turns out to be a real loser. A loser with big names will draw many more than a loser with unrecognizable names.

A similar process appears to underlie the growth of social networks as well. In a recent study, sociologists Fredrik Liljeros and Christofer Edling of Stockholm University, working with a team of physicists from

Boston University, looked at the links of sexual contact between 2,810 randomly selected individuals in Sweden in 1996.[6] If acquaintance is a fairly loosely defined relationship, the existence or nonexistence of a sexual link is not (Bill Clinton's arguments notwithstanding). In this social context, Liljeros and his colleagues found the same structure as the World Wide Web or the Internet: a small world, with a few people dominating the number of sexual links within the community, the signature power-law pattern for the distribution of people according to their number of sexual partners.

In the sexual context, these are the people that Malcolm Gladwell referred to in his book *The Tipping Point* as the "connectors," those socially prolific few who tie an entire social network together. In Stanley Milgram's original letters experiment, for example, the letters that made it from Nebraska or Kansas to his stockbrocker friend in Boston did not make the final step to their destination from just anywhere. Of the letters that arrived at the stockbroker's home, a full two-thirds were sent there by a friend who was a clothing merchant, a person Milgram referred to as Mr. Jacobs. Others arrived at the stockbroker's office, and the bulk of these came from just two other men, whom Milgram called Mr. Brown and Mr. Jones. Gladwell rightly emphasized the strangeness of this condensation of letters into just a few hands: "Think of it. Dozens of people, chosen at random from a large Midwestern city, send out letters independently. Some go through college acquaintances. Some send their letters to relatives. Some send them to old workmates. Everyone has a different strategy. Yet in the end, when all of those separate and idiosyncratic chains were completed, half the letters ended up in the hands of Jacobs, Jones and Brown."[7] These people are the connectors, the hubs of the social world, with thousands of friends and acquaintances, far more than most people have. When rumors, news about a job opening, or letters from a strange experiment make their way through a social network, they tend to pass through the connectors—much as airline flights have a great tendency to pass through hub cities such as Atlanta and Chicago.

But Gladwell did not investigate where such connectors come from. In the case of the sexual network, for example, why isn't sexual activity centered about an average or typical number of sexual links? Select three thousand people at random, and no one will be ten times taller than another; their heights will fall around some well-defined average. Measure how fast people can run, or how much weight they can lift, and the distribution works similarly. But sexual activity is not like this. You might put the prolific performance of the connectors down to special skills bestowed at birth or in early childhood, and yet, as Liljeros and his colleagues suggested, there may be another explanation: "Plausible explanations for the structure of the sexual-contact network described here include increased skill in acquiring new partners as the number of previous partners grows . . . and the motivation to have many new partners to sustain self image. . . . Evidently, in sexual contact networks, as in other 'scale free' networks, the 'rich get richer.'"[8] The term "scale-free" is a technical reference to the power-law or fat-tail distribution we have seen so many times for network elements, according to how many links they have.[9]

Aspects of basic social psychology might also support the process of preferential attachment. In a famous study in 1952, for example, the social psychologist Solomon Asch had subjects in groups of six study a line drawn on a page. They were then to say which of three lines on an adjacent page was the same length. Five of the subjects in each group were working with Asch and purposefully gave, out loud, the same incorrect response. Asch found that by hearing the others, the sixth subject—the true focus of the experiment—was frequently swayed to give the same wrong answer, conforming to the group rather than trusting his or her own senses. A few of the sixth subjects even reported later that they had perceived the lines differently.[10]

These findings show how easily our decisions, even our perceptions, can be influenced. If you think of our hypothetical furniture maker as Asch's sixth subject, then you can see that there is a certain psychological force pushing him or her toward choosing links to the already popular Web sites. After all, the popularity of a site is public testimony to the fact that a large number of others have already proclaimed, through their actions, that "this site is worth visiting."

This effect also has a link to another famous idea of social psychology known as *groupthink*. In the 1970s, social psychologist Irving Janus explored the way groups of people come to make decisions, and concluded that often group dynamics limit the group's ability to legitimately consider alternative options. To reduce the psychic discomfort of disagreement, members of a group strive to find a consensus, and once a rough consensus is reached, it becomes difficult for dissenters to voice their ideas. They may keep quiet, not wanting to make waves. "Concurrence-seeking," as Janus wrote, "becomes so dominant in a cohesive group that it tends to override realistic appraisal of alternative courses of action."[11]

In the context of the Web, the creator of a site chooses its links with absolute freedom. But making a choice and putting the site online is roughly analogous to expressing a personal opinion in one of Janus's groups. Similar social-psychological dynamics may be at work here, influencing people to link up with the same sites over and over again.

THE OLD BOYS' NETWORKS

YOU MIGHT THINK scientists are immune from such herdlike behavior, since they are meant to be driven by their unbiased quest for the truth. But in studies of how scientists choose whom to work with, researchers again have found the rich getting richer. Consider the network of scientists linked by having coauthored a paper together. For this network there are excellent data recording when each link was established—this being the date when the paper was published—and how many previous collaborators each scientist had beforehand. Therefore, it is possible to test the rich-get-richer idea explicitly, by going into the bitter detailed history and checking on the choices that scientists actually made. When establishing a new collaboration, did scientists tend preferentially to link up with scientists who had already collaborated with a large number of others?

Several researchers have explored the literature in physics, neuroscience, and medicine in just this way, and in each case the numbers reveal that indeed the rich get richer. For example, as Mark Newman of the Santa Fe Institute concluded from the statistics of one study involving Medline, a large database maintained by the National Institutes of Health for papers published in biology and medicine, "The probability of a particular scientist acquiring new collaborators increases with the number of his or her past collaborators."[12]

Researchers have put the growth of the Internet and the World Wide Web under similar historical scrutiny, and in every case backed up the intuitive picture we have been describing. All speculation about social psychology aside, these results are a mathematical confirmation of a universally influential effect.[13]

Not surprisingly, the tendency for name recognition to influence decisions also has a major influence in the business community, determining who sits on what corporate board, for example. It has been a matter of controversy for nearly a century that the boards of directors of America's major corporations are heavily "interlocked"—that is, they are linked to one another by individuals who sit on more than one board. For those of a more conspiratorial bent, this is ominous news, for it suggests that the boards of different corporations are sharing information, fixing prices, and scheming and conspiring to wield immense power across the economic and political landscape.

From a mathematical point of view, there is certainly good reason for some eyebrows to be raised. Two years ago, Gerald Davis and colleagues from the business school at the University of Michigan looked into the network of interconnections between corporate boards and discovered yet another small world. Suppose we think of two boards as being linked if one person sits on both. Alternatively, we could think of two businessmen as being linked if they sit on the same board together. These are two separately conceived but closely related networks. Davis and his colleagues studied both networks, finding similar results for each. They concluded that "corporate America is overseen by a network of individuals who—to a great extent—know each other or have acquaintances in common. On average, any two of the 6724 Fortune 1000 directors we studied can be connected by 4.6 links, and any two of the 813 boards are 3.7 degrees distant." This implies that the boards of the major U.S. corporations are tied together socially into one immense web of corporate government. Indeed, as Davis and his colleagues added ironically, "It is literally true that an especially contagious airborne virus would spread quite rapidly through the corporate elite."

Davis and others have shown that the ties within this network have important influences within the business community. Corporations whose boards include a high number of officers from banking institutions, for example, tend to borrow money more frequently, suggesting that the connection to bankers may color their decisions. In general, links to other corporations help to spread information or attitudes from one board to another, and help corporations to keep their fingers on the pulse of current thinking in many different branches of the economy. An automaker might well benefit from having links to the boards of major oil producers or steel manufacturers.

However, this elite mafia of the business world is perhaps not quite the conspiratorial network it might seem. Early in the twentieth century, critics of capitalism such as Vladimir Lenin in Russia insisted that

large banks were the evil organizing forces behind this network of corporate elite. But recent studies suggest no undue influence of the
bankers. Indeed, Davis and colleagues concluded that the emergence of
the small-world structure is hardly surprising since it turns up in
almost every natural network studied. As they suggested, "It is difficult
to think of a policy that would eliminate the small world property of
the corporate elite, short of banning interlocks altogether."[14]

Where does this small-world structure come from? In earlier work,
Davis found a suggestive clue—that directors who are already heavily
interlocked are more likely to be chosen for new board positions. They
are made more attractive, that is, by virtue of already being popular and
sitting on a number of other boards. To begin with, a director with links
to several boards will more likely have access to a broad range of useful
information from different industries, a wider breadth of ideas. In
addition, members of boards of directors are chosen not only for their
advice but also to enhance the reputation of the firm in the eyes of
potential investors. As sociologist Mark Mizruchi of the University of
Michigan suggests, "By appointing individuals with ties to other important organizations, the firm signals to potential investors that it is a
legitimate enterprise worthy of support."[15]

All of these findings suggest that the small-world structure within
the corporate world emerges for the very same reason that it does in so
many other networks—as a result of the rich getting richer. At many
levels and in numerous settings, this simple process appears to be the
inanimate architect of the small world.

THE FLAVORS OF SMALL

THE PRINCIPAL MESSAGE of our story so far is that small worlds are
almost everywhere. They arise in such tremendously different settings
that deep general principles must lie behind their emergence. At the
outset, we were facing a mystery: how, in a world of six billion individuals, can any two possibly be linked by no more than six degrees of separation? Oddly enough, we now have two distinct answers.

On the one hand, Duncan Watts and Steve Strogatz, building on the
clues of Mark Granovetter, pointed out how a few long-distance bridges
can make all the difference within a network. This is clearly a point of

central importance, and yet the networks of Watts and Strogatz lack
two striking features of many real-world networks. To begin with, they
do not grow. Every complex network in the real world has come to be
through a history of growth: the Internet and World Wide Web over
several decades, ecological food webs over millions of years, and so on.
The networks of Watts and Strogatz also lack the vital ingredient of history while at the same time lacking connectors, the few rare elements
that possess a disproportionate share of all the links.

In terms of the way links get shared among the elements of a network, networks of the Watts and Strogatz variety are "egalitarian," as
the links are distributed more or less equally. In contrast, highly connected hubs or connectors dominate the networks of Albert and
Barabási. The historical mechanism of the rich getting richer leads
without fail to connectors, who, by virtue of having so many links, naturally play a role similar to that of Granovetter's bridges. They have a
more complex form than simple long-distance links, but nonetheless
bring together regions of a network that would otherwise be quite distant. In striking contrast to small-world networks of the egalitarian
kind, these networks with hubs might be better described as "aristocratic," as only a handful of elements possess most of the network's
links.

So there are, it seems, two flavors of small: egalitarian networks in
which all the elements have roughly the same number of links, and aristocratic networks characterized by spectacular disparity. In this chapter
and previous chapters we have taken a close look at the Internet and the
World Wide Web, at the networks of sexual contacts between people, of
scientific papers linked by citations and of scientists linked by having
coauthored papers, and of words linked by appearing next to one
another in English sentences. In each of these aristocratic networks
there are hubs or connectors, presumably the consequence of the rich
getting richer.

But for other small-world networks, this is not the case. The neural
network of the nematode worm *Caenorhabditis elegans*, for example,
has no connectors, as each neuron is linked to roughly fourteen others.
The same egalitarian character seems to describe the neural network of
the human brain, as well as transportation networks of many kinds,
including the webs of roads and railways that cover the continents. In
the case of the U.S. electrical power grid—the transportation network

for electrical energy within the United States—each generator, transformer, or substation links up with roughly three others, and again, there is a conspicuous lack of highly linked connectors.

What are we to make of all this? Why have some networks come to have one flavor and others another? Is there any purpose here? Or is it simply an accident?

8

COSTS AND CONSEQUENCES

Everything is what it is because it got that way.
—*D'Arcy Wentworth Thomson*[1]

OVER THE PAST two centuries, the distance travelled each day by the average person has grown by a factor of ten thousand. According to one estimate, an average person in 1800 traveled no more than about 50 meters each day. Many stayed within and around the home, or worked in the fields, and most of those who worked in the towns and cities lived there. There was little commuting back then. Nowadays we travel an average of 50 kilometers each day.[2]

At first it was the horse, the canals, and the great ocean-going vessels that made us more mobile. Later it was the railway and automobile. Today millions pour into and out of the major cities on a daily basis, while more than twenty million flights crisscross the globe each year on the routes of the flourishing air transportation network. During the year 2000, Chicago's O'Hare International Airport alone pushed more than seventy-two million passengers through its gates, while the Hartsfield Atlanta International Airport handled even more, an incredible eighty million, representing nearly twice the entire population of the United Kingdom.

Not surprisingly, the air network was beginning to buckle under the strain a couple of years ago. In the United States, the summer of 2000 was the worst in history for air traffic delays. At O'Hare, some 4,600 flights were cancelled and another 57,000 delayed, numbers typical of major airports all over the country. In the preceding five years, the number of delays exceeding forty-five minutes had doubled, and serious delays of this kind were affecting 10 percent of all flights.[3] Aircraft that could fly 600 miles an hour were being forced to average only 250 between Washington, D.C., and New York City.[4] Here as elsewhere, the problem was simply too much traffic; if planes flew at normal speed, they would have accumulated over the airports in swarms of circling aircraft.

In the spring of 2001, the U.S. Congress held a series of hearings to get to the root of the problem, and listened as George L. Donohue, professor of systems engineering and operations research at George Mason University in Virginia, painted a sobering picture. "The US hub and spoke air transportation system," he said, "is approaching a serious capacity crisis." Overall, he pointed out, the network was then running at 58 percent of maximum capacity, and would be at 70 percent in just ten years.[5] In air-operations lingo "maximum capacity" means the absolute, full limit, and statistics compiled by the Federal Aviation Administration show that serious delays kick in anytime an airport runs over 50 percent of its maximum capacity. An airport is like a supermarket, where the Saturday-morning crowd makes it almost impossible to maneuver even though a lot more shoppers could still be packed into the open spaces.

In year 2000, the three largest U.S. airports, in Atlanta, Chicago, and Los Angeles, were already running at over 80 percent of maximum capacity. And that was during good weather. In bad weather, air traffic controllers had no choice but to increase the spacing between planes for safety, and under these conditions, nearly half of all major U.S. airports were operating above their maximum capacity. The consequences were as unsurprising as they are annoying: delays and cancellations all over the country whenever Mother Nature stirred up a little bad weather.

Things were no better elsewhere. As chief executive of British Airways Rod Eddington commented at the time, "Congestion in the skies and on the ground in Britain is becoming critical. Air traffic staff do a magnificent job and safely. But there is a finite limit to the numbers of aircraft they can allow into a sector at any given time. . . . Heathrow is groaning at the seams."[6] Two years ago, the world air network was rapidly reaching its limits, as airport capacity worldwide was failing to keep up with an explosive growth in demand. What was to be done? Build more runways? Unfortunately, studies reveal that once an airport has three or four runways, as the larger ones do already, building more runways yields diminishing returns.[7] There may be more places to take off or land, but aircraft still have to get to and from the runways, and planes end up mired in traffic jams on the ground.

Of course, the situation has changed entirely, at least temporarily, in the aftermath of the September 11 terrorist attacks in the United States. With airport security heightened worldwide and many passengers afraid to fly, the number of flights has fallen dramatically and airlines have laid off workers by the thousands. Airport congestion is the least of the airlines' problems. Nevertheless, it is worth looking back at the issue of congestion, and not only because the problem will again take center stage when air traffic rises to its former level. At the end of chapter 7, we raised the question of what causes the difference between the two kinds of small-world networks—the egalitarian and aristocratic networks, as we have called them. Air traffic congestion may seem to have little to do with a subtle theoretical issue of this sort, and yet when it comes to gaining a deeper understanding of small-world networks, and especially of their origins, the case of the airport turns out to be particularly illuminating.

FROM RAGS TO RICHES . . . AND BACK TO RAGS

IF YOU HAVE a Web site and want to link it to Yahoo or Amazon.com, go ahead. The fact that these are principle hubs of the World Wide Web to which innumerable other sites already link does not present any obstacles. There are no laws or Web guidelines that limit the number of links, and so these other links do not "get in the way" of your own. There is nothing in the Web to hamper the richest sites getting still richer, and as we have seen, this process leads inexorably to a small-world network of the aristocratic kind, with a few elements acting as superconnected hubs.

In the past, this is how the network of flights linking airports has grown, as airlines have sought centralized operations in major hubs offering easy links to many destinations. O'Hare advertises itself as offering "more connections to more cities, more often than any other airport in the world," which may be true enough. It is also a well-designed and managed airport; readers of *Business Traveller* magazine voted it "the best airport in America" two years in a row. So, in principle, an airline locating its operations at O'Hare ought to gain lots of advantages, and O'Hare should have an edge in the competition for new airlines and nonstop routes. But over the past decade and perhaps even longer, O'Hare and the other major airports also have offered more congestion, more delays, and more cancellations. So unlike the situation with Yahoo, Amazon.com, or other Web hubs, the links to O'Hare are indeed getting in the way.

The situation suggests that for some time the richest have had trouble getting still richer, and the busiest airports have not enjoyed their true advantage in the competition to attract new flights and new airlines. Congestion seems to be getting in the way of still further growth. What, if anything, does this have to do with the two different kinds of small-world networks? Late in 1999, physicist Luís Amaral and colleagues from Boston University by coincidence were studying the system of air connections as a network, and they noticed something peculiar. Despite all the usual talk of the largest airports as hubs, the mathematics did not bear this out. A close study of the network actually places it as a small-world network of the egalitarian kind, that is, as one conspicuously lacking in hubs. Any two airports can be linked by no more than five flights along common nonstop routes, but the distribution of airports according to the number of links they have does not follow the fat-tail pattern. Superconnected hubs are actually far more rare than they are in the World Wide Web or the Internet.

Seeking an explanation, Amaral and his colleagues took a few steps back and imagined a network growing by the basic rich-get-richer pattern. As they knew, Réka Albert and Albert-László Barabási had proved six months before that if even the most heavily linked elements in a network never face any difficulty in adding still others, the result will always be an aristocratic network in which a few hubs dominate. But what if the richest eventually begin to struggle under their burden? Tweaking the basic picture of a growing network ever so slightly, so as to bring such a factor into play, Amaral and the other physicists worked out the consequences in a series of calculations and computer experiments. Their results fit in beautifully with the statistics for the airports and in retrospect, make perfect intuitive sense. As the network grows, the rich get richer for a time, and hubs do emerge. But eventually the most highly linked elements begin to lose their advantage in gathering new links. Hence, the less connected sooner or later begin to catch up to the more highly connected, and the network then becomes more egalitarian, each element having roughly the same number of links.[8]

This point may seem somewhat technical, a detail for the specialists, and yet there is an important "take-home" message: small-world networks of the egalitarian variety, à la Watts and Strogatz, are far more than mere mathematical curiosities. Like the aristocratic networks of the World Wide Web or the Internet, small-world networks of the egalitarian kind can emerge from a simple process of history and growth. Whenever limitations or costs eventually come into play to impede the richest getting still richer, then a small-world network becomes more egalitarian, as seems to be the case with the airports and a number of other real-world networks.

As we have seen, for instance, roughly three links emerge from every element in the electrical grid in the United States, and roughly fourteen sprout from each neuron in the neural network of the simple worm *Caenorhabditis elegans*. These small-world networks are of the egalitarian kind, and we can imagine why. At some point it becomes technically difficult and just too expensive to make still more links to one electrical substation. These elements are not Web pages to which other pages can connect merely by including a hypertext link. Electrical substations live in the real, physical world where there is only so much space to use in making new connections with a bulky apparatus. When substations get too crowded, it becomes easier and more efficient to simply link up somewhere else.

In the case of air travel, there may have been true hubs in the past, before the larger airports became so burdened, but now the move toward egalitarianism may be picking up steam. A number of smaller secondary airports are beginning to compete with the larger hubs. After all, with more free space they can offer more punctual handling of aircraft. Regional airlines that fly shorter distances are also becoming more profitable and popular. According to one estimate, the number of smaller aircraft serving regional airports will double in the next two decades.[9] For now, many of these airlines are still serving the hub airports, connecting people to major urban centers from less populated areas. But congestion is beginning to drive the smaller craft away from the hubs.

Whatever the eventual effect on ticket prices and the availability of seats, the air transportation network is starting to look far more like the web of neurons in the brain than the Internet's network of computers. All are small worlds but of somewhat different sorts, and with a little insight from small-world theory, we have a good idea why.

THE SENSES OF SMALL

THIS STRIKING INSIGHT regarding how limitations influence the growth of a network brings us to a kind of unified theory of small-world networks. It blends the thinking of Albert and Barabási together with that of Watts and Strogatz, and fits their ideas together like pieces of a puzzle into a larger and more coherent picture. On the one hand, the rich-get-richer mechanism leads inevitably to small-world networks, as if they were dictated by an architectural law of nature. Nevertheless, limitations and constraints sometimes get in the way and leave their telltale traces on the final form. Still, the similarities between the two kinds of networks are probably more important than their differences. The small-world character persists in either case, the differences being more like the differentiating details of the delicate friezes on a great cathedral, rather than the gross structural points that would make one cathedral recognizably Gothic and another Romanesque.

It is fair to say that a host of puzzling questions still remain, as is hardly surprising in a field of research that is even now barely four years old. How does the architecture of a network change when elements get taken away? What if elements grow old and die? Or what if some of the internal links within a network get rewired? Then again, what if the pace of growth changes, sometimes being faster and other times slower? Technical questions such as these are keeping a growing body of researchers occupied. But there are also questions of a deeper sort, answers to which would have more far-reaching implications.

The ultimate value of all this research lies not only in identifying a new kind of architecture, in working out how to describe it and to detect its subtle nuances, and in appreciating how it differs from all previous conceptions of networks. More important is what this discovery teaches us about the world, especially in a practical sense. It is one thing to see that many seemingly unrelated networks are actually similar; it is another to use that insight to make computer networks that run faster and crash less frequently, organizations that make better decisions, or airports that face less congestion. In a few areas, small-world research is already making important practical contributions.

As just one example, computer scientists until recently were at a complete loss as to how to model the growing structure of the Internet. This issue is considerably more serious than it may seem. Before researchers can develop the next generation of Internet "protocols," the common operating rules that permit computers within the net to talk to one another and exchange information efficiently, they need to be certain that the protocols will work properly on the Internet as it really is, both now and in the future. As it turns out, protocols designed for one network architecture sometimes perform poorly on another, triggering, for example, information logjams that cause computers to crash. Hence, getting the architecture right is crucial, and proposed protocols have to be tested on realistic mock-ups of the Internet. Fortunately, testing is now possible since researchers have discovered the true Internet architecture and have a means for generating networks of the very same sort.[10]

This is just one minor example. The small-world discovery and other ideas now growing out of it represent one of the first great successes of the theory of complexity. For if the small-world architecture arises through the work of what is an almost inexorable law of physics, it also possesses a range of remarkable properties. As scores of researchers are now finding, this architecture confers distinct advantages on any network that might possess it. Its advantages are by no means completely or even partially understood, but in tracking them down, researchers are beginning to make some impressive headway.

CYBER-THREATS

THE UNITED STATES has the largest economy in the world and the most powerful military. It is the world's only remaining superpower, and yet the country clearly faces many serious threats. The terrorist attacks of September 11 and the subsequent terror of anthrax have totally dispelled any illusions about U.S. invulnerability. Together with the other nations of the West, the United States faces serious threats from forces located both inside and outside of its borders. And yet despite the shock of September 11, many had seen the trouble looming. As long ago as 1996, for example, President Bill Clinton underlined the urgency of such new threats, which are, at least in part, wrapped up with the double-edged nature of the world's revolutionary new computer technology: "The threat to our open and free society from the organized forces of terrorism, international crime and drug trafficking is greater as the technological revolution, which holds such promise, also empowers these destructive forces with novel means to challenge our security. These threats to our security have no respect for boundaries and it is clear that American security in the 21st century will be determined by the success of our response to forces that operate within as well as beyond our territory."[11]

Everything from the postal service and power grid to the banking system and air traffic control now depends crucially on integrated computer networks, and it is natural to wonder if such networks are safe from attacks by terrorists or foreign countries, not to mention natural disasters. Two years later, the White House again responded to growing recognition of the vulnerability of what has come to be called the United States' "critical infrastructure." A presidential directive of May 1998 acknowledged that "our economy is increasingly reliant upon interdependent and cyber-supported infrastructures and non-traditional attacks on our infrastructure and information systems may be capable of significantly harming both our military power and our economy."[12]

Clinton's concerns are illustrated not only by the global drama that has unfolded from September 11, but by many other alarming facts and incidents. In the first half of 1999, computer viruses cost the U.S. economy some seven billion dollars. In May of 2000, the "I LOVE YOU" virus spread to seventy-eight million computers worldwide within just four days, ultimately causing some ten billion dollars worth of damage. Almost every day the newspapers carry stories about cyber-crime of one sort or another, as funds go missing from banks or teenage hackers play tricks. In one spectacular and well-known incident, concerted "denial of service" attacks in February of 2000 against Amazon.com, CNN, eBay, and Yahoo brought each Web site to its knees for several hours. In these attacks, known as *smurf attacks*, perpetrators sent out a huge number of Internet packets to third-party computers on the Internet, each packet carrying a false return address—the address of one of the intended targets. When these packets arrived at their destinations, the computers receiving them automatically sent return packets back toward the target. Hundreds of millions arrived all at once, overwhelming the local transmission lines and effectively wiping these sites off the Web.

You might think that the sprawling computer network of the U.S. Department of Defense would be far more secure than any commercial enterprise. After all, this network would be at the center of much of the information coordination during any U.S. military effort. But in 1997, in a test of the security of the defense network, hackers working for the National Security Agency were able to break into some thirty-six separate defense computer systems, where they simulated turning off sections of the U.S. power grid. They also gained access to the electronic systems of a Navy cruiser at sea.[13]

Fortunately, most attacks so far have been the work of private hackers probing the network in an unsophisticated way, either for personal gain or in the interest of pure mischief. In one fairly typical incident, computer expert Steve Gibson of Gibson Research Corporation waged a concerted battle for several months to keep his Web site up and running in the face of attacks by a thirteen-year-old who called himself "Wicked." "You can't stop us," Wicked claimed in a message that he posted anonymously, "because we are better than you, plain and simple."[14] There is no guarantee, however, that cyber-crime in the future will remain mostly cyber-annoyance. In testimony before Congress in 1996, the director of the Central Intelligence Agency, John Deutch, pointed out that U.S. intelligence services have good evidence "that a number of countries around the world are developing the doctrine, strategies, and tools to conduct information attacks. . . . I am convinced that there is a growing awareness around the world that advanced societies—especially the U.S.—are increasingly dependent on open, and potentially vulnerable information systems."[15]

In 1999, a report of the RAND Corporation on the threat to the U.S. information infrastructure warned that extrapolating the recent history of uncoordinated attacks by independent hackers into the future could be dangerous: "We must consider what a well-funded, determined, sophisticated adversary might accomplish, and how—and at least prepare for such a contingency as a worst case."[16]

Others have come to the same conclusion. According to a report produced at the U.S. Army War College, the casual hacker is not likely to cause any truly significant damage to U.S. national security. More worrying is the well-funded intelligence service of a foreign power, which might be able to plan and execute a concerted attack. The report concluded that "a well-coordinated, state-sponsored attack against the Federal Aviation Administration could cripple the nation's airline industry and cause planes to crash. Likewise, an attack on financial institutions could disrupt the banking system and cripple the stock market thereby destabilizing the economy. State-sponsored attacks could disrupt entire communities, states, or even the entire nation. . . . The infrastructure may well be resilient to the perturbations of man-made cyber-disasters, but it is not immune from the effects of a well-coordinated attack."[17]

These conclusions are not based on mathematics but on careful perusal of how the network of critical infrastructure might be attacked intelligently, and the safeguards now in place. But the crucial role of coordination in any successful network attack emerges even more clearly from the perspective of small-world networks.

FAILING GRACEFULLY

HOW WOULD A network such as the Internet actually fare under attack if, for example, some information terrorist started knocking off the computers on which it depends? Does the Internet's small-world architecture make it any more or less resilient than it might otherwise be? In the winter of 1999–2000, Réka Albert, Hawoong Jeong, and Albert-László Barabási reckoned that they were probably as well placed as anyone to attempt an answer. After all, having worked out many of the details of the Internet's actual architecture, they could run simulated attacks against it, and also against various other kinds of networks, and see what happens. Which would stand up and which would fall apart? When people talk about network safety and resilience, they generally emphasize the role of redundancy, of having several elements able to perform the same basic tasks so that if one gets wiped out, one of the others can step into its place. Redundancy makes perfect sense—no army unit would ever head into battle with only one soldier knowing how to operate the radio or a key piece of weaponry. Even so, Albert and her two colleagues revealed that redundancy is not enough.

Suppose we imagine two different Internets, one a small-world network of the aristocratic kind, much like the real Internet, and the other a purely random network having the same total number of computers and links between them. Both of these networks are redundant: while knocking out any one computer will destroy some paths within the network, others will remain to take up the slack. But there are subtle differences in how these networks fare when the destruction becomes more severe.

To begin with, Albert, Jeong, and Barabási looked at the effects of haphazard failures, the kind of trouble a network might face from the occasional breakdown of some of its computers or from an uncoordinated and unsophisticated attack. With the random Internet, they began knocking out computers one after another, selecting each target by chance, while monitoring the network's diameter—its number of degrees of separation—which offers a rough indication of how well connected it is. To no one's surprise, the results showed that the diameter of the network grew steadily as various elements were wiped out. By the time one of every twenty elements had been removed, for instance, the network diameter had grown by 12 percent.

As the destruction proceeded, the situation became dramatically worse. By the time 28 percent of the elements had been destroyed, this randomly constructed network had completely disintegrated into a collection of small, isolated subnetworks. Even though 72 percent of the computers were still up and running, the network was so fragmented that each computer could communicate with only a few others. If an army were relying on this network, it would be in a sorry state. Random networks, despite their redundancy, fall apart quite quickly in the face of an uncoordinated attack.

The three physicists next repeated the same attack against a small-world network of the aristocratic kind, and found good news: networks with the actual structure of the Internet fare remarkably better. Even when 5 percent of the elements were knocked out, the diameter of the network was unchanged. Moreover, this network fell apart gracefully under the attack, and never suffered a catastrophic disintegration. Even with nearly half of all the nodes removed, those that remained were still sewn together into one integrated whole. Instead of suddenly shattering, bits of the network slowly flaked off like chips from a rock while the rest stayed together.

It is the highly connected hubs that account for the difference between the two networks, as the hubs act as a kind of glue within the network. Since an uncoordinated attack targets elements at random, it almost always knocks out unimportant elements with few links, while missing the hubs. In this way, the small-world architecture makes a network resilient against random failure or unsophisticated attack.[18] This finding presumably explains why the Internet, despite the continual failure of routers and other hardware elements, never collapses as a whole. There is, however, a rather alarming corollary to this conclusion that echoes the fears of military theorists regarding the dangers of coordinated attack. As it turns out, the very feature that makes a small-world network safe from random failure could be its Achilles' heel in the face of an intelligent assault.

In further simulations, Albert, Jeong, and Barabási studied how their two test Internets would fare if the most highly connected hub computers were wiped out first—this being a better strategy from an attacker's point of view. For a network such as the real Internet, the simulations revealed that knocking out only 1 percent of the network elements pushes the diameter up by 12 percent, and destroying just 5 percent doubles it. As far as the network's wholeness or integrity is concerned, the destruction of 18 percent of the elements serves to splinter the network almost entirely into a collection of tiny fragments. Under coordinated attack, a small-world network is a sitting duck—in fact, the random network has the advantages.[19]

What lessons can we draw from this? Military analysts already appreciate the obvious point that particularly important elements in any network require special protection. As suggested in the RAND report, "Some information infrastructures are so essential that they should be given special attention, perhaps in the form of special hardening, redundancy, rapid recovery, or other protection or recovery mechanisms."[20] To some extent, the U.S. government is already following this advice by using firewalls; data encryption as well as brute physical barriers to protect the principle telephone exchanges; major elements in the power grid and pipeline control systems; key air traffic control sites; and so on. But the small-world perspective brings home just how crucial such extra protection might be. Without it, a few well-conceived strikes might suffice to fragment the information infrastructure into hundreds of small, isolated, and nonfunctioning pieces. Building redundancy into a network is not nearly enough—more subtle features of the architecture can have important effects.

Of course, looking at the basic architecture of a network represents only one part of any effort to protect it against failure and attack. In the case of the U.S. information infrastructure, some analysts suggest that the United States needs to adopt an aggressive policy of preemptive strikes against "cyber-warriors" in hostile countries. Others believe, however, that complex networks can be safeguarded in a more sophisticated way by learning to copy the defense mechanisms of living things. Much as the immune system recognizes foreign invaders, such defenses would have the ability to detect an attack quickly, and would respond by isolating the damage to one part of the network. It would also be able to adapt "on the fly" by rerouting information through the network, redistributing tasks, and even mounting immediate counterstrikes.

These are ideas of the not-too-distant future, and yet this biological way of thinking seems particularly apt and raises intriguing ideas about other possible uses for the small-world theory. After all, as we know, the architecture of the Internet shares deep similarities with that of the biochemistry of the living cell. What advantages, if any, does the cell gain from the small-world trick? How does the resilience and vulnerability of the small-world architecture play itself out in the biological world? And might this teach us anything about how to attack microbes more intelligently? As it turns out, Albert-László Barabási and his very busy colleagues have been pondering these questions too.

BIO-WARFARE

EVEN AN ORGANISM as simple as the single-celled bacterium *Escherichia coli* has nearly five thousand genes that participate with one another in supercomplex biochemical networks. Loosely speaking, each gene is a small strip of DNA that can be either "on" or "off." Any gene, when on, teams up with other cellular apparatus to produce protein molecules of various types that make up the components of the cell's membrane, act as sensors on the outer surface of the bacterium as it searches for food or tries to avoid danger, or carry signals from one part of the creature's innards to another. Genes also produce proteins that act to turn other genes on or off, and it is these cascading control signals—gene affecting protein, affecting gene protein, and so on down the line—that in delicate combination orchestrate cell division and bacterial reproduction, help maintain the chemical balance inside the cell, and guide the bacterium as it "swims" toward sources of food. The immense complexity of this chemical symphony beggars the imagination and yet somehow carries out subtle and finely tuned functions more sophisticated than anything man-made technology can match.

In terms of basic network architecture, however, as we saw briefly in an earlier chapter, there is a curious link between this network, the Internet, and the World Wide Web. To picture the cell as a network, think of each important molecule as an element in the network, with links connecting molecules that can participate with one another in the chemical reactions that make the cell live. In the immediate aftermath of their work on the vulnerability of the Internet, Barabási and his colleagues turned their attention to these cellular networks, and with biologists from Northwestern University studied forty-three different organisms, including the bacterium *E. coli* and stretching across the spectrum of different kinds of life. In every organism, these researchers found the very same pattern: a small-world network of the aristocratic kind. A few molecules play the role of highly connected hubs and take part in far more reactions than most others. Molecules such as ATP, adenosine triphosphate, crucial in providing for the cell's basic energy needs, comprise one such hub and form part of the functional core of the cellular network.

In any aristocratic small-world network, the hubs are so crucial that when they are destroyed, the network breaks into pieces. When we think about the national infrastructure, as we discussed, this reliance on hubs is an Achilles' heel for the Internet and other information networks, as an attack against the hubs can be devastating. But in the biological context of the cell, the aristocratic structure may be a reason for hope—it may help in the search for crucial new drugs.

Much as the developed world is racing to protect its network infrastructure from attack, it is also racing to find new ways to attack the biological and fungal agents that cause disease, many of which have in recent years developed immunity against the traditional treatments. Most of the antibiotics used today were derived from agents first used nearly half a century ago, and in that time bacteria have grown increasingly clever in finding ways to foil their action. According to the U.S. Institute of Medicine, "Microbes that once were easily controlled by antimicrobial drugs are, more and more often, causing infections that no longer respond with treatment to these drugs."[21]

The bacterium *Staphylococcus aureus*, for example, which causes food poisoning, is typically treated with an antibacterial agent known as methicillin. But in some hospitals, as many as 40 percent of food poisoning cases involve strains that are resistant to methicillin. These patients have to be treated instead with another drug called vancomycin. The trouble is, using vancomycin so frequently gives *S. aureus* plenty of chances to evolve resistance to this newer drug as well. In the United States and in Japan, at least five separate strains have been identified as having developed partial resistance already.[22]

Identifying the deeper architecture of a bacterium or yeast may not help to solve this problem immediately. Nevertheless, it may take us a significant step toward building a greater appreciation for how cells function and where they are most vulnerable. For example, there are good reasons why the network diameter, or number of degrees of separation of a cellular network, should be small. After all, the diameter is directly related to how many chemical reactions have to take place in a chain before a significant event in one part of the network triggers meaningful effects in others. A bacterium, for instance, may swim into an area high in glucose, its favorite molecular food. A protein sensor on the bacterium may recognize the good news and respond by boosting the concentration of other molecules inside the bacterium, so setting off a chain reaction of multiple stages intended to produce a healthy supply of the molecules needed to digest glucose.

This chain of events in a bacterium is like Pavlov's response in a dog, and to be efficient, it has to take place quickly, which is possible given the small-world character of the network. With just a few cascading reactions, in just a few steps in the chemical network, the right work gets done. On the other hand, anything that would significantly disrupt the network architecture and make the diameter much larger could have dire consequences. This possibility raises the obvious question, what would happen to the network diameter if this or that molecule were somehow obliterated? Given the close link between cellular networks and the Internet, the answer is fairly obvious. If molecules were knocked out at random by an unintelligent attack, then there should be little increase in the diameter, just as with the Internet. In contrast, if molecules were attacked more intelligently, we would expect a drastic and sudden fragmentation of the biochemical network.

This scenario is purely theoretical, but real experiments bear it out. Biologists can selectively cause genetic mutations that lead to the deletion of one specific molecule in the network—an enzyme, for example, that acts to catalyze certain other reactions. Such experiments reveal that in the bacterium *E. coli*, a large number of catalytic enzymes can be removed without much effect on the organism's ability to survive; on the other hand, removal of a crucial few has devastating effects.[23]

The same pattern is found in more complicated creatures too, such as the yeast *Saccharomyces cerevisiae*, more commonly known as brewers' or bakers' yeast. Biologists mapped out the complete genome of this organism in 1996, finding sixteen chromosomes and about 6,200 genes. But the physical functioning of the organism's chemistry is actually more easily apparent at the level of proteins, the molecules these genes produce. In yet another study spurred by their new insights into complex networks, Barabási's group in the spring of 2001 teamed with Sean Mason of the Department of Pathology at Northwestern University in Chicago to explore the effects of knocking out specific proteins in the biochemical network of this yeast (Figure 17). Methodically, the group deleted each protein one by one in experiments and looked at how the "connectivity" of that protein—its number of links in the network—matched up with the consequences of its removal.

The results were striking. More than 90 percent of the proteins in the network had five or fewer links, and only about one in five of these was essential to the yeast's continued survival. With these removed, the yeast could still function by adapting its remaining network. In contrast, less than 0.7 percent of the proteins were hubs having more than fifteen

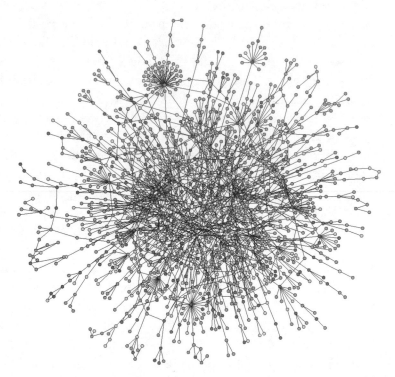

Figure 17. A diagram showing the network of interactions between the various proteins in the yeast *Saccharomyces cerevisiae,* more commonly known as brewer's or baker's yeast. (Image courtesy of Hawoong Jeong, reprinted by permission.)

links. A single deletion of any one of these was lethal in two out of every three cases. As the researchers concluded, "Highly connected proteins with a central role in the network's architecture are three times more likely to be essential than proteins with only a small number of links."[24]

These findings clearly lead to a deeper understanding of cellular architecture at a basic level, and of the relative importance of the network's various parts. But they may prove practically useful as well. The network perspective suggests that bacteria and other microbes can be hit hardest by striking at the proteins most highly linked within the biochemical network. Like the central command centers in any army, these hubs keep the entire network up and functioning.

SMALL-WORLD THINKING

THE RESEARCH REVEALS just a few of the more obvious and important insights now tumbling out of the small-world perspective. It suggests not final answers but promising new ideas and approaches toward a number of serious problems now confronting those who struggle with the workings of complex networks. On a larger scale, does the small-world structure make our ecosystems stable? Or is it instead the signature of a network that is dangerously vulnerable to sudden, catastrophic collapse? How does the structure of our social networks affect the spread of diseases, such as AIDS? And can understanding the social architecture help us to combat such diseases?

In the remaining chapters, we will explore the lessons and implications of the emerging small-world theory of networks, lessons that range from the management of ecosystems to the understanding of the workings of the human brain. As we already know but will see in more detail shortly, networks of this sort—whether of the egalitarian or aristocratic variety—possess a power and adaptability far beyond what they might have otherwise. Moving beyond the small-world theory, we will also see more generally how an appreciation for the way organization can emerge in complex networks of all kinds, small worlds and not so small worlds alike, is changing the face of science.

9

THE TANGLED WEB

A science is any discipline in which a fool of this generation can go beyond
the point reached by the genius of the last generation.
—*Max Gluckman*[1]

IN JULY OF 2001, a pro-whaling institute of the Japanese government
took out half-page advertisements in numerous Japanese and international newspapers. "Whales are increasing as fish stocks decline," it
cried, "whales are threatening our fisheries!" The advertisement was
timed to coincide with the July meeting in London of the International
Whaling Commission, where delegates from more than forty countries
were set to debate issues surrounding the worldwide moratorium on
whaling, now fifteen years old. Japan hopes to cancel the moratorium
and to resume commercial whaling.

Under the rules of the moratorium, countries can still slaughter as
many whales as they like for the purposes of "scientific research." Japan
has taken full advantage of this loophole, killing some four hundred
minke whales each year, which end up as whale bacon in Japanese
restaurants and as diced blubber in the markets. As part of this
"research," fishermen cut open the stomachs of captured whales and
not surprisingly, they have discovered that whales eat fish. Hence the
claim that it is the whales who are to blame for dwindling fish populations. A Japanese man who runs a struggling whale meat–processing
plant paraphrased the argument: "The whales eat the same small fish
our fisherman catch. . . . Before the whaling moratorium, we had coex-
istence. . . . Now, because of the whaling ban, the minkes do whatever
they want, and our fish catch is decreasing."[2] For the Japanese fishing

industry, this argument offers a convenient conclusion, but only by standing reality on its head. As a Greenpeace spokesman quite rightly put it, "This is like blaming woodpeckers for deforestation," for the scientific evidence is simply overwhelming that it is not the whales but commercial fishing that is to blame for the devastation of marine ecosystems worldwide.

Last year, for example, an international team of nineteen marine ecologists completed an exhaustive historical study of marine life in coastal ecosystems ranging from coral reefs and tropical seagrass beds to river estuaries and continental shelves. How do fish populations today compare to those in the past? In every case, the record shows a precipitous decline in fish numbers with the onset of modern methods of industrial fishing.[3] As the authors of the study concluded, "Overfishing of large vertebrates and shellfish was the first major human disturbance to all coastal ecosystems. . . . Everywhere, the magnitude of losses was enormous in terms of biomass and abundance of large animals that are now effectively absent. . . ."[4]

The situation has become critical in the past few decades. The world's marine fish catch peaked at about 85 million metric tons in 1989 and has been declining ever since, with stocks of Atlantic cod reaching historic lows, while haddock and other species have been declared commercially extinct. Thriving food webs that were stable for millions of years have been radically altered in the past twenty years, and according to the United Nations Food and Agriculture Organization (FAO), nearly three-fourths of the world's commercially important marine fish stocks are now fully fished, overexploited, or depleted.[5]

So it is hard to see how anyone can sensibly blame gluttonous whales, especially since the fish and whales have been swimming side by side in the oceans for millions of years, and large-scale, industrialized fishing is the new factor in the equation.

But the Japanese argument is in fact even weaker than it seems. Suppose we agreed with the Japanese delegate to the International Whaling Commission, who sees the minke whale as the "cockroach of the sea." Suppose we were even willing to exterminate the whales in an attempt to catch a few more delicious fish. Would the wholesale eradication of whales really boost the fisheries' take? Is the marine ecosystem so simple that we can pull a lever in one place and anticipate the effect in another? As we will see, the trouble with this way of thinking is that it refuses to acknowledge the true complexity of our ecosystems. Ecological reality is decidedly not so simple, as the fishing industry should have learned long ago.

BATTERED COD

IN THE MID-1980S, the number of cod in the Northwest Atlantic Ocean began to plummet. The sudden decline in the population came as a shock to the Canadian fishing industry and the government, which set up a task force of respected scientists to look into the problem. The group, led by Leslie Harris, then president of Memorial University in St. John's, Newfoundland, ultimately came to a conclusion that proved to be somewhat unpopular with the politicians: "Failure to take appropriate steps to reduce current levels of fishing mortality will most probably lead to a significant continuing decline in the spawning population."[6] The Canadian trade minister at the time said he would be "demented" to follow the advice of Harris's advisory group, and rejected it without consideration, complaining that "Harris doesn't have to deal with the economic, social and cultural effects of reduced quotas. I do." Unfortunately for the trade minister, nature is not put off by fine words. Despite their best efforts, by 1992 Canadian fishing vessels were falling far short of their quotas and for a very simple reason: the Atlantic cod population had collapsed, and there were very few fish left to catch. Canadian fishermen had to face up to the "economic, social and cultural effects" after all, and with a vengeance, as the entire cod fishing industry was shut down.[7]

Even then, the government refused to acknowledge any responsibility for the disaster and instead blamed all sorts of other factors. The real cause was not overfishing, they insisted, but the intrusive fishing habits of European fishing boats and, oh yes, the voracious appetites of North Atlantic harp seals who were eating all the fish. Further scientific assessment again pointed to a very different conclusion. In 1994, two scientists working for the Canadian Department of Fisheries and Oceans concluded that European fishermen and seals had absolutely nothing to do with the problem and that simple overfishing was the key.[8] A year later, one of these scientists, Ransom Myers, was quoted in a prominent Canadian newspaper: "What happened to the East coast fish stocks had nothing to do with the environment, nothing to do with seals, it was simply overfishing."[9] Apparently this was an inconvenient bit of information for the government, and Myers, a government employee, was reprimanded for speaking out.

Meanwhile, despite what its own scientists said, and in the face of the concerted scientific opposition of marine biologists worldwide,[10] the Canadian government stuck to what they insisted was the "commonsense" conclusion: seals eat cod, and so they must be at the root of the problem. Throughout the latter half of the 1990s, government-organized hunting expeditions slaughtered nearly half a million harp seals each year, allegedly to help the cod population recover, which it didn't.

What the government failed to appreciate—or found politically expedient to ignore—is that North Atlantic harp seals eat, in addition to cod, other fish such as capelin, hake, herring, and halibut, close to 150 other species in all, many of which are direct competitors of cod.[11] Halibut eat cod, as do seabirds, squid, and sculpin, another fish species eaten by seals. Taking the rest of the oceanic food web into account, we can begin to appreciate the staggering complexity of the situation (Figure 18). A decrease in the number of seals would directly affect at least 150 other species in the web. Changes in the numbers of these in turn would affect countless others, sending millions of competing chain reactions rippling through the food web.

Indeed, considering chains involving only eight species or less, ecologists estimate that more than ten million distinct chains of cause and effect would link the seal to the cod.[12] In the face of this overwhelming complexity, it is clearly not possible to foresee the ultimate effect of killing seals on the numbers of some commercial fish. With fewer seals off the Canadian coast, the number of halibut and sculpin might grow, and since they both eat cod, there may well end up being *fewer* cod than before.

This kind of reasoning applies just as well to whales, which not only eat commercial fish but often eat the predators of those fish. So no one can honestly claim to know what the effect of resumed commercial whaling might be—other than a dangerous depletion of already endangered species. Will there be more fish? Or fewer fish? Politics and public-relations propaganda to one side, it is anyone's guess.

Is there something special about oceanic food webs that makes them so complex? Hardly. To probe the structure of a food web, ecologists will often sequester a small area and remove one particular predator to

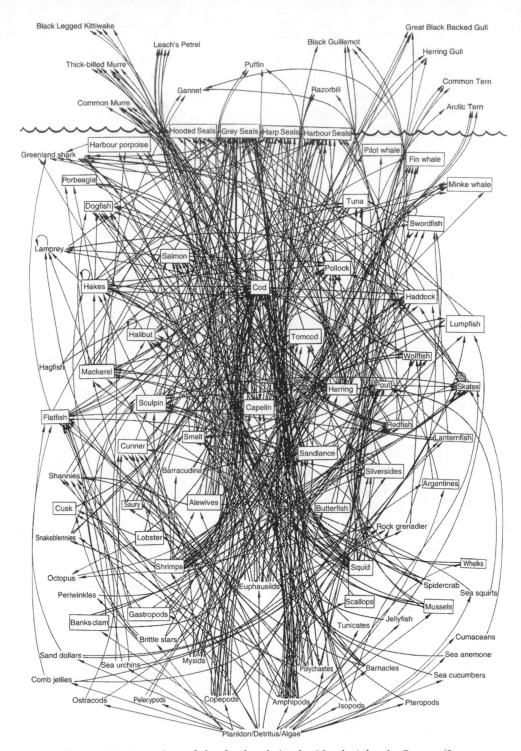

Figure 18. A portion of the food web in the North Atlantic Ocean. (Image courtesy of David Lavigne, reprinted by permission.)

see what effects this has on its principal prey. You might expect the results to be fairly predictable: remove the predator, and its prey should flourish. Surely the Canadian and Japanese fisheries would agree. But in 1988, ecologist Peter Yodzis of the University of Guelph in Canada compiled the results of experiments of this sort carried out in thirteen different ecological communities from around the world. He found that the effect of removing a predator, even on its most obvious prey, was generally unpredictable. The trouble lies with the great number and complexity of the indirect pathways that link species together.[13]

These examples begin to illustrate the immense complexity of our ecosystems and give immediate cause for grave concern. Biologists estimate that the rate of species extinctions worldwide is a thousand times greater now than it was before humans walked the earth, and if extinctions continue at the present rate, fully one-fourth of all life-forms will be obliterated in fifty years. What effect will this devastation have on the global ecosystem, the living network that supports *our* continued existence as well? The leaders of governments and large corporations worldwide find it convenient to think like the Canadian trade minister, suggesting that worries about the ecosystem are overstated and that it would be "demented" to carry out any reforms that are not politically or economically popular. But what is the science of ecosystem stability?

A food web is a network much like a social community or the Internet, with species linked to one another in a tangled pattern that, to the untrained eye, appears completely lacking in any overall organization or plan. As we know, however, order is often hidden, and different kinds of organization lead to different properties for the network as a whole. Can the global ecological network withstand the human pressure now put against it? Or is it likely to fall apart? With a bit of mathematics, researchers are closing in on answers to these questions, and in the process are reaching some alarming conclusions regarding the fragile nature of the web of life.

THE STABILITY OF COMPLEXITY?

BEFORE THE 1970S, the commonsense view in ecology was that a network of interacting species was generally more stable the more rich and complex it was. An ecosystem with a huge number of species of different kinds, in this view, would be more stable and less susceptible to sudden catastrophic change than would an ecosystem made of only a few species. One of the first ecologists to argue this way, Charles Elton of Oxford University, claimed that simple communities are "more easily upset than richer ones; that is, more subject to destructive oscillations in populations, and more vulnerable to invasions."[14] Elton pointed to a number of examples. Small islands, for instance, tend to have relatively few species, and habitats of this sort are much more vulnerable to invaders than are habitats on continents where there are many species. In a similar vein, pest invasions and outbreaks tend to plague natural lands less frequently than cultivated or planted lands, where human activity has greatly simplified the ecological community.

This pattern fits in well with our understanding of the living world as a rich tapestry of interacting creatures and plants, and our association of a loss of this richness with damage to the system. But is it really true?

In the early 1970s, Australian ecologist Robert May applied some mathematics to the problem and came up with a surprise. May looked at the stability of schematic networks of interacting species. In his models, each species fit into an ecosystem as a predator or prey, and he adjusted the complexity of the network by adding more species or taking them away or by increasing or decreasing the number of links between them. The idea was to study the stability of these "model ecosystems" by seeing whether, after being disturbed, they would be able to withstand the disturbance and settle back to normality.

From a theoretical point of view, this was far more sophisticated than anything ecologists had done earlier, and it pointed to a conclusion in direct conflict with ecologists' experience. May found that as the number of interacting species grew, it generally became less likely that the network would be able to settle back down after being perturbed. Instead, with increasing complexity, a disturbance became more likely to shake the network to pieces, triggering large and uncontrollable fluctuations in species' numbers, with many going extinct. In contrast, simpler and less complex networks were more stable. Networks of this sort could withstand the interference of the environment or the invasion of some new species without severe upheaval.[15]

Given the seeming inevitability of May's mathematical reasoning, ecologists could not help but be impressed. Maybe complexity was not so important after all! Ecologist Ransom Myers now refers to this as the beginning of a notable phase in the history of ecologists' debate over ecosystem stability, a phase in which, as he put it, "Ecologists, terrified of complex math, go into a coma for 20 years." But in recent years, the field has entered into the next phase, in Myers's words: "Ecologists recover and begin thinking again." In fact, at least one ecologist had begun thinking again even a bit earlier.

Peter Yodzis started out his scientific career in the early 1970s at the University of Bern in Switzerland. As a physicist working in general relativity, he wrote papers filled with mathematics and bearing titles such as "On the Expansion of Closed Universes." Ten years later, however, Yodzis turned toward theoretical ecology, now producing papers with titles such as "Black Bear Population Dynamics." Given his mathematical background, Yodzis was in a far better situation than most ecologists to scrutinize the meaning and implications of May's work.

Among other observations, Yodzis noticed that May had used random graphs to build his networks. Like Paul Erdös in pure mathematics or the early social scientists, he had connected the elements haphazardly. Are real ecosystems truly random networks? Yodzis did not know, but to find out he began studying the literature and compiling data for a number of real ecosystems, trying to build a better picture of how actual food webs are put together. He also included information on how strongly various species interacted with one another. In a social network, the bonds between good friends are not the same as those between weak acquaintances. And so it is with species: A predator may rely solely on one prey as its food, and so "interact" very strongly and frequently with it. Or it may prey on one hundred different species and have relatively weak and infrequent interactions with all of them.

Yodzis put all the information he had into his models and then, like May, studied how they stood up to disturbances. The results were intriguing. In contrast to random networks, these more realistic food webs were both highly complex *and* highly stable. They were well able to take considerable shocks, or the extinction of one species, without going to pieces. These results provided only a hint to be sure, but they suggested that May's random networks were missing some crucial ingredients. Some kinds of complex networks—indeed, the kind that look more like real ecosystems—are stable after all.[16]

Complexity, then, is sometimes a good attribute, and recent experiments in the field have demonstrated Yodzis's point even more convincingly. In a painstaking series of experiments during the 1980s and early 1990s, a team of researchers led by David Tilman of the University of Michigan divided four fields of grassland in the Cedar Creek Natural

History Area in Minnesota into 207 plots, and studied these plots to determine the complexity of each. More specifically, they measured the "diversity"—the number of different species—and at the same time monitored how vigorously the total biomass within each plot fluctuated. The biomass is the measurement you would get if you gathered up everything living in the plot, dried it out, and then weighed it. It offers a picture of how well the ecosystem is doing in terms of producing living stuff.

Tilman and his colleagues' goal was to see if there was any link between the complexity within a plot and the fluctuations of its biomass, the idea being that larger fluctuations would be associated with less stability and vice versa. The researchers discovered a striking trend across all the plots: the greater the complexity and the greater the number of species, the weaker the fluctuations in biomass.[17] Conclusion: more complex networks tend to fluctuate less and are more stable than simpler networks, in direct contrast with Robert May's mathematics.

These findings suggest that there may be some architectural laws for the world's ecosystems. Rather than being merely random networks of species linked together haphazardly, there may be a good deal of sense lurking in the detailed patterns of the connections between species. Food webs have rightly been called "the roadmaps through Darwin's entangled bank,"[18] and in the past few years, researchers have begun to follow these road maps to uncover deeper principles of ecosystem design. Stability, it turns out, seems to be closely associated with complexity after all, and especially with certain kinds of "weak ties" that link species together into richly connected food webs.

THE SECURITY OF WEAK TIES

IN TERMS OF its novelty and explanatory power, Robert May's attempt to understand the stability of ecosystems had everything going for it. With mathematics, ecologists seemed poised to answer general questions about ecosystems once and for all. And yet May's approach suffered from some subtle shortcomings—in particular, a somewhat inadequate notion of the idea of "stability."

May started from the idea of an ecosystem as being in a condition of perpetual balance. Think of a chair. A chair is stable if you can give it a gentle nudge with your foot and it remains standing as before. This is how May envisioned the stability of an ecosystem. A precipitous change in climate might push the ecosystem away from its natural, balanced condition, boosting the numbers of some species at the expense of others. But what happens afterward? Do the populations soon settle back to what they were before? If so, then the ecosystem, whether real or mathematical, is stable. On the other hand, if populations do not settle down, if the numbers of the various species change from what they were before, or if the numbers of some soar while others go extinct, then the ecosystem is unstable.

May's way of thinking about stability is perfectly legitimate, and yet it is certainly not the only way and perhaps not the most appropriate. In 1998, ecologists Kevin McCann, Alan Hastings, and Gary Huxel, then all at the University of California at Davis, took an important step forward by thinking about ecosystem stability from a subtly different and less restrictive point of view.

In any real ecosystem, the numbers of various species fluctuate from year to year. This is completely normal. The number of foxes or rabbits this year is not exactly what it was last year or the year before. Fluctuations do not imply that the ecosystem is unstable, merely that it is subject to change. Even so, natural fluctuations of this sort make May's idea of stability hard to apply. His definition of *stable* requires the numbers of all species, after a disturbance, to settle down to what they were before. For the chair example, this definition would require each leg, after the nudge, to settle back to rest in *exactly* the same position it was before. Otherwise, it would be considered unstable.

But is it really so important that everything remain unchanged? Can't something be stable even if it is altered a bit by a disturbance? A chair, for example, might end standing but six inches away from its initial resting point. That is a legitimate kind of stability as well. Or, in an ecosystem, the populations of every species might end up being different from what they were before, and yet the community might still hang together and persist in an altered but healthy condition. McCann and his colleagues suggested that this would be a better way of thinking about ecosystem stability. After all, the ultimate question is whether the ecosystem falls apart or not. The issue is whether it has the capability to absorb a shock and hang together, to persist, even if in an altered condition.

This change in approach may seem an unimportant shifting of the rules, but it is actually much more. Thinking of food webs in terms of their ability to persist helps to illuminate some of the great driving forces of instability in any ecosytem, those that stir up violent fluctuations in the numbers of various organisms. As McCann and his colleagues pointed out, not all interactions between species are alike. Some are far stronger than others. And it is the stronger interactions that lead to trouble.

Species interact by eating one another or by competing for the same prey or habitat. If a predator eats just one other species, it will do so frequently, having no other options. The interaction between these two species will be strong. Conversely, if a predator feeds on fifteen different prey, it may eat each species only occasionally. In this case, it will have relatively weak interactions with these species. Now suppose that owing to a recent change in climate or some other chance factor, the number of a predator's only prey has been severely depleted. This predator will now have difficulty finding food, and yet it has no option—it must continue to seek its prey even though its numbers are so small, driving this species even closer to extinction. When this happens, the population of predators may then fall precipitously as well. The strong link between these species sets up the possibility for dangerous fluctuations in their populations.

In direct contrast, McCann and his colleagues argue, weaker links can save the day.[19] Consider a predator with fifteen different prey, for example. If the numbers of one of these species become very low, for whatever reason, the natural response of the predator would not be to drive those numbers lower still but to shift its attention to another species. After all, being more numerous, any of the other fourteen prey would now be easier to catch. As a result of this shift in attention, the predator would continue to find food, while the prey in danger of extinction could revive its numbers. In this way, weak links between species act to take the wind out of dangerous fluctuations. They are the natural pressure valves of ecological communities.

From this point of view, weak links between species play a special role in tying an ecological community together, which sounds strangely reminiscent of Mark Granovetter's point about the strength of weak ties in our social networks. In itself, this observation is not terribly profound. And yet it brings important questions into focus: Which are the weak links in an ecosystem, and which are the strong? Is the link with social networks anything more than accidental? Thinking carefully about small-world networks suggests some intriguing possibilities.

Suppose, for example, that a food web was much like a social network. Suppose it was a small-world network of the aristocratic kind, with most species linked to only a handful of others, but a few "connector" species with links to a huge number of other organisms. Where would the weak and strong links fall in such a network? In a social context, we already know the answer. If someone has five thousand "friends," after all, they clearly cannot all be close friends. No one has the time and social energy to maintain such strong bonds with a great many people. Consequently, the superconnected few should be linked to others mostly by weak links, while those with few links to others should be connected by strong links. By analogy, the same would be true in an ecosystem. If a species has a huge number of links to others, most of these will be weak links. An organism can only eat so much, for example, and if it feeds off 150 different species, it is likely to feed infrequently on each.

So if ecosystems were small worlds of the aristocratic kind, they would naturally be dominated by the weak ties of the superconnected few. From these few they would derive the natural stability that goes with a preponderance of weak ties. Of course, this is only conjecture. If true, then the superconnected few would be the lynchpins of the community, the most precious and invaluable species that prop up the entire community. But do such species really exist? Are ecosystems really like social networks?

TWO DEGREES OF SEPARATION

WHEN ECOLOGISTS THINK about food webs, they do not always have in mind creatures devouring one another in the dark oceanic depths, or lions stalking terrified wildebeest across African plains. In the ecological research literature, in fact, some of the most informative food webs carry names such as "Wet Stump Hole, Alabama" or "Dog Carcass, Costa Rica," inconspicuous yet thriving ecosystems in miniature that make up for their lack of obvious drama by the rich communities of insects and microbes they support. Other food webs of scientific note are little more than plots of ordinary land, such as an area of 97 hectares in southern England that is maintained by biologists from Imperial College, London. Part of Imperial College's Silwood Park research area, this ecosystem represents a complex food web of 154 species associated with Scotch broom, a plant of the pea family that colonizes grasslands and grows to about 2 meters in height.

It is hard to imagine anything more boring than 97 hectares of pea plants. But what gets ecologists excited about this plot of land is that they understand its food web in intricate detail. Over several decades, researchers have mapped out the links between all 154 species and can draw an exact diagram of the food web. And with this knowledge, they can gain insight into crucial questions. How many species-to-species links does it really take to link any two organisms in some chain of cause and effect? How many links does it take for a disturbance in one part of a food web to trigger effects in other distant parts? Last year, physicist Ricard Solé and ecologist José Montoya set out to answer these questions in the case of Silwood Park, and the answers they found are anything but reassuring.

The species found in this 97-hectare plot include flies, bugs, spiders, and beetles. There are rabbits and birds, as well as bacteria and fungi,[20] and one principle plant, the Scotch broom. Using a computer, Solé and Montoya took each possible pair of species and calculated the number of links in the shortest species-to-species path connecting them. With more than 150 species, the number of degrees of separation in this food web could be quite large, perhaps as high as fifty or sixty. But Solé and Montoya found an extremely small world: between any bird and beetle, spider or bacterium, the number of degrees of separation was typically only two or three. The tapestry of life at Silwood Park is woven from a truly dense cloth of interconnections.[21]

Perhaps Silwood Park is unusual? To find out, Solé and Montoya extended their study to a community of 182 species in the freshwater ecosystem of Little Rock Lake in northern Wisconsin. They also studied a food web involving 134 species from the Ythan River estuary, located about 20 kilometers north of Aberdeen in the United Kingdom. Meanwhile, another team of researchers led by Richard Williams of San Francisco State University put to the test seven other distinct food webs sampled from ecosystems globally. Each of these studies found exactly the same thing: small worlds with only two or three degrees of separation.

Of course, one plot of farmland is not the global ecosystem. And there are certainly more than two steps from a species of woodpecker in Illinois to a shrimp in the South China Sea. Even so, whales and many species of fish populate the oceans as a whole, and numerous birds migrate between the continents, providing all-important long-distance links that tie the biological world together. For the global ecosystem, the number of degrees of separation may not be two, but it is probably not much higher than ten. In view of what we have seen earlier in the case of social networks or the Internet, this discovery is hardly surprising. And yet it contrasts strikingly with traditional ecological theory, in which the "distance" between species ought to grow in rough proportion to the size of a food web.

The small-world architecture prevents such proportional growth in distance, and keeps the biological world sewn tightly together. Not only will the culling of one species affect every species it eats, competes with, or is eaten by, but also it will send out fingers of influence that in a few steps will touch every last species in the global ecosystem. As Williams and his colleagues concluded, "Most species within a food web can be thought of as 'local' to each other and existing in surprisingly 'small worlds' where species can potentially interact with other species through at least one short trophic chain. . . . This suggests that the effects of adding, removing or altering species will propagate both widely and rapidly throughout large complex communities."[22]

So ecological communities are small worlds. They are also characterized by connectors. Taking their network analysis further, Solé and Montoya looked at the number of links each species has to others in the food webs of Silwood Park, Little Rock Lake, and the Ythan River estuary. For each web, they counted up how many species were linked to two others, how many to three others, and so on, and then made a graph. The curve in each case turns out to follow the power-law or fat-tail pattern that we have seen so many times: if 100 species have two links, then only 50 have four links, only 25 have eight links, and so on. The number of species having a certain number of links falls off by a constant factor each time you double the number of links. The exact numbers are not so important, but this pattern is the signature of an aristocratic network, with the dominance of a few superconnected hubs, species that possess a disproportionate fraction of all the links in the entire web.

Hence ecosystems are indeed like social networks, and the small-world perspective offers a deeper understanding of the ecological importance of weak ties. Any hub or connector species has a huge number of links to other species. As a result, most of these links will be weak links; the two species interact infrequently. Now, since the network's connectors possess such a large share of all the links in the entire web, it follows that most links in the web will be weak links. In other words, the preponderance of weak links in an ecosystem emerges directly from its small-world architecture. This architecture by itself provides the biological pressure valves that help to redistribute stress and prevent one species from wiping out another by uncontrolled predation or competition.

In this sense, the aristocratic, small-world structure is a natural source of security and stability in the world's ecosystems. This insight, however, is not wholly comforting. As we found in the case of the Internet, this architecture also has an Achilles' heel.

KEYSTONES TO COLLAPSE

HALF OF THE tropical forests, where two-thirds of all species find their habitat, have now been logged or burned to clear land for human development, with another 1 million square kilometers disappearing every five to ten years.[23] How dangerous is this loss of species diversity for the world as a whole? If healthy ecosystems are small worlds characterized by connectors, and weak links provide their stability, then the global depletion of species numbers is a truly alarming prospect. For as species continue to disappear, the remaining species in our ecosystems will come to interact more strongly, if only by simple arithmetic.

If a predator preys on only six species where before it preyed on ten, its interactions with the six will be stronger, and ecosystem stability will only suffer. Simplified ecological communities also may be more vulnerable to invasion by foreign species. So if species continue to vanish, we can expect the future to be increasingly grim. As Kevin McCann sees it, "We should expect an increase in frequency of successful invaders as well as an increase in their impact as our ecosystems become simplified. The lessons for conservation are obvious: (1) if we wish to preserve an ecosystem and its component species then we are best to proceed as if each species is sacred; and (2) species removals (that is, extinction) or species additions (that is, invasions) can, and eventually will, invoke major shifts in community structure and dynamics."[24]

While the small-world effect may be beneficial in so far as it binds us together socially into one world, it is rather troubling from an ecological point of view. Since no species is ever far from any other, it is unlikely that any species anywhere on the planet will long remain unaffected by human activity. Perhaps there are none left even now. At the very least, this cautions doubly against the careless removal of species for resource "engineering" purposes, as advocated by the Canadian or Japanese fishing industries, for example.

What's more, the consequences of removing just one connector species can be especially dramatic, as a huge number of weak stabilizing links goes with it. Ecologists have long talked about "keystone" species, crucial organisms the removal of which might bring the web of life tumbling down like a house of cards. From the small-world perspective, the connectors look like the keystones. And Solé and Montoya have demonstrated just how crucial their preservation may be. Suppose you take an ecosystem and begin removing species. Slowly but surely, the food web should fall apart. But how will it fall apart? And which species are most crucial in holding it together?

To find out, Solé and Montoya took one more look at the food webs of Silwood Park, Little Rock Lake, and the Ythan River estuary. They considered two ways in which an ecosystem might come under attack. On the one hand, an ecological community under pressure—from human activity, climate change, or what have you—might suffer the extinction of a few species more or less at random. Using a computer to mimic the loss of random species from the food web, the researchers found, encouragingly, that real communities stand up relatively well. As species disappear, the "diameter" of the food web grows very slowly, and the total number of species remaining in one central, fully connected web of life decreases gradually. This is the good news: ecosystems endure fairly well when a few species are removed at random. A fair fraction of the species can even be removed without the entire web falling to pieces.

But there is disconcerting news as well. Suppose species are not removed at random, but that the most highly connected species get knocked out first. In this case, as Solé and Montoya discovered, ecological disaster ensues quickly. Indeed, removing even 20 percent of the most highly connected species fragments the web almost entirely, splintering it into many tiny pieces. As the web falls apart, the disintegration triggers many "secondary extinctions" as well, as some species lose all of their connections to others and become totally isolated. These simulations underline the obvious point: the true keystones in an ecological community are the most highly connected species, the hubs of the network.

These keystones are the ecological control centers, so to speak, and clearly the most important targets for preservation. In the past, ecologists have suspected that large predators would tend to be the keystones in an ecosystem, but this does not seem to be true. In their three ecosystems, Solé and Montoya found that the highly connected keystones were often inconspicuous organisms in the middle of the food chain or were sometimes basic plants at the very bottom of the web. In other cases, they were major predators. There appear to be no hard and fast rules for determining which kinds of species are likely to be keystones. This lack of rules may make it more difficult for ecologists to identify the most important species, and yet it also suggests the best way to proceed. Identifying keystones means studying the network architecture and seeing which species are the connectors, the lynchpins of the living fabric.

What Solé and Montoya achieved in their computer, human activity on a global level is achieving in the real world: the methodical dismantling of the world's ecosystems. With almost no theoretical understanding at all, we can only hope that the ecosystem has the ability to withstand this assault without collapsing. We know next to nothing, and what we do know is worrying. The small-world perspective on ecosystems is only a beginning but at least offers a few insights that may help us to mitigate the destruction, by identifying keystone species and offering a more realistic picture of the dense connections between species. Understanding the stability of an ecosystem, and learning how to manage more intelligently our interactions with the other living things on Earth, appears to require as much understanding of networks as knowledge of specific organisms.

What makes the network perspective so powerful is that it reaches beyond the details of this or that setting, beyond computers, airports, or organisms, to identify deep and influential principles of organization at work "behind the scenes." In nature, it is often not the character of the individual parts of a network that matter most, but the overall order—or lack of order—dwelling within it. This discovery is hardly new. Physicists have known for well over a century that the molecules in ice and water are identical. When a lake turns into a skating rink in winter, it reflects not a change in the molecules themselves but a subtle alteration in the pattern of molecular organization, a network property that no study of a single water molecule could ever foresee.

This network perspective, however, has truly come into its own in the past decade. Physicists in particular have entered into a new stage of their science and have come to realize that physics is not only about physics anymore, about liquids, gases, electromagnetic fields, and physical stuff in all its forms. At a deeper level, physics is really about organization—it is an exploration of the laws of pure form.

10

TIPPING POINTS

I'm firmly convinced that not only a great deal, but every kind, of intellectual activity is a disease.

—*Fyodor Dostoyevsky*[1]

O N THE EVENING of April 28, 1938, special agents of the NKVD, the Soviet Directorate for State Security, arrested three physicists of the famous Institute for Physical Problems in Moscow. The NKVD had been watching Yuri Rumer, Moissey Koretz, and Lev Landau for two years, and according to files made available by the KGB in the late 1980s, the police suspected the former two of being members of "a counterrevolutionary wrecking organization headed by Landau." The arrest occurred at the height of the Stalinist purge of every conceivable political enemy or potential troublemaker. In the previous two years, an estimated 850,000 members of the Communist Party—more than a third of its total membership—and nearly ten million people in all had been either shot or shipped off to concentration camps, never to be seen again.

In this case, the NKVD had ample evidence against the conspirators. Agents had obtained a copy of an anti-Soviet leaflet that Rumer, Koretz, and Landau were preparing to distribute on May Day,[2] and its wording made their deaths almost certain:

Comrades!

The great cause of the October revolution has been evilly betrayed. . . . Millions of innocent people are thrown in prison, and no one knows when his own turn will be. . . .

Don't you see, comrades, that Stalin's clique accomplished a fascist coup! Socialism remains only on the pages of the newspapers that are terminally wrapped in lies. Stalin, with his rabid hatred of genuine socialism, has become like Hitler and Mussolini. To save his power Stalin destroys the country and makes it an easy prey for the beastly German fascism. . . .

The proletariat of our country that had overthrown the power of the tsar and the capitalists will be able to overthrow a fascist dictator and his clique.

Long live the May day, the day of struggle for socialism!

The Antifascist Worker's Party

Ordinarily, anyone connected with such a leaflet would have been shot. For mysterious reasons, however, the three physicists were spared. Koretz was sent to the Gulag—the Soviet system of forced labor camps in Siberia and the far north—where he spent twenty years, surviving to return to Moscow in 1958. The state sent Rumer to a special Gulag prison devoted to scientific and engineering work, where he spent the next ten years. Meanwhile, Landau, the alleged ring leader, was hauled off to Lubyanka prison in downtown Moscow, where six months later he signed a full confession, admitting that "at the beginning of 1937, we came to the conclusion that the Party had degenerated and that the Soviet government no longer acted in the interests of workers but in the interests of a small ruling group, that the interests of the country demanded the overthrow of the existing government." Remarkably, Landau was not executed but remained in prison, where fortune smiled upon him for a second time six months later. Early in 1939, internationally famous physicist Pyotr Kapitsa wrote to the Soviet prime minister, Vyacheslav Molotov, saying that he had made a curious discovery in his laboratory and that no physicist besides Landau could possibly explain it. Landau was soon released.

Had the Soviet security apparatus decided otherwise, physics today would be very different. Landau explained Kapitsa's discovery within a few months,[3] and over the next three decades left his mark on virtually every area of physics, from astrophysics and cosmology to the study of magnetic materials. Landau also invented a revolutionary new theory of phase transitions, a theory of how substances of all kinds change their forms. It could be solid ice melting to liquid in a gin and tonic, or dry ice turning to smoky vapor on a theatrical set. In essence, Landau's theory was a theory of networks.

In trying to explain and understand such phase transitions, Landau's theory pushed the substances themselves to the background and brought the more abstract elements of molecular and atomic organization into focus. As a result of Landau's visionary ideas, much of modern physics now is not really about matter at all, but about discovering the laws of form in networks of interacting things—not only atoms and molecules, but also bacteria and people. The small-world idea is just part of this larger theory of networks, which rests on a deep truth first suggested by Landau: a collection of interacting elements often has properties that depend in no crucial way on the nature of the elements themselves.

We will return to Landau's idea in more detail later in this chapter. But first it will be useful to explore a few of the important issues that Landau's ideas can help us to understand. A book of recent years that seems to have fired the public imagination is Malcolm Gladwell's *The Tipping Point*, an exploration of the notion that ideas, rumors, waves of crime, and many other kinds of influence can spread through a society in much the same way as a virus. Ultimately, the book is about the way influences spread through a network of interacting things—in this case, people. It is hard to imagine that theoretical physics could lend support to a way of thinking that is triggering a revolution in advertising and marketing. But as we will see, the conclusions of *The Tipping Point* find a comfortable home within the modern extensions of Landau's theory.

HOW IDEAS ACQUIRE PEOPLE

THE CENTRAL IDEA of *The Tipping Point* is that tiny and apparently insignificant changes can often have consequences out of all proportion to themselves, and that this accounts for the fact that sweeping changes often rise up out of nowhere to transform industries, communities, and nations. The nub of the idea, as Gladwell put it, is that "the best way to understand the dramatic transformation of unknown books into best sellers, or the rise of teenage smoking, or the phenomenon of word of mouth or any number of other mysterious changes that mark everyday life is to think of them as epidemics. Ideas and products and messages and behaviours spread just like viruses do."[4] If true, he argued, this insight would help explain sudden and far-reaching social changes of innumerable kinds. Consider a few examples.

As of the beginning of 1994, a traditional brand of brushed-suede American shoes, Hush Puppies, were selling about 30,000 pairs a year. Hush Puppies were about as hip as hair curlers, and as far as American tastes were concerned, this style seemed to be quite literally worn out. Everyone thought so at least, until 1995, when sales suddenly went through the roof. In that year the manufacturer sold more than 430,000 pairs, and the next year even more. The company's executives were as surprised as anyone, for nothing they had done could account for the sudden upswing.

Around the same time, a similarly inexplicable transformation swept over New York City. In 1992, the city was plagued by 2,154 murders and 626,182 serious crimes, and residents feared to walk the streets after dark. Some particularly rough areas were even off limits to the police. But again, for reasons that were impossible to pinpoint, the situation began changing in a hurry. By 1997, the number of murders had dropped by 64 percent, and the number of serious crimes was cut in half. The police, not surprisingly, claimed that more officers and new tactics had done the trick. But city governments announce initiatives to get tough on crime every year and never achieve such startling results. What was different in this case?

These are the kinds of transformations that Gladwell wants to explain. Think of "grunge" music sweeping out of Seattle and across America in the late 1980s, ethnic cleansing spreading like a deadly plague through the Balkans in the early 1990s, or the wave of support that, at least for a time, brought Ross Perot into serious contention as a third-party candidate in the U.S. presidential election of 1992.

It is no secret that ideas and kinds of behavior can be contagious. Indeed, the notion goes back a long way. As the financier Bernard Baruch once suggested, "All economic movements, by their very nature, are motivated by crowd psychology. Without due recognition of crowd-thinking . . . our theories of economics leave much to be desired. . . . It has always seemed to me that the periodic madnesses which afflict mankind must reflect some deeply rooted trait in human nature. . . . It is a force wholly impalpable. . . . yet, knowledge of it is necessary to right judgement on passing events."[5] That "deeply rooted trait in human nature" is little more than a susceptibility to influence and a predilection toward imitation. In the 1630s, a mania for tulips swept through Holland, and as the demand fed on itself, the price of common varieties soared by a factor of more than twenty, before crashing precipitously. At one point, a single tulip bulb could fetch the same price as 12 acres of prime farmland.[6] When it comes to money and investment, history is a long string of speculative bubbles driven by the infectiousness of human behavior. Take the dramatic rise and fall of the "dot com" Internet stocks just a couple years ago. These are bandwagon effects driven by the wavelike spreading of beliefs.

Twenty-five years ago, the evolutionary biologist Richard Dawkins suggested that there might be a genetic element to the logic of spreading ideas. Just as genes pass from generation to generation, Dawkins suggested that ideas—he called them "memes"—may do the same: "Examples of memes are tunes, ideas, catch-phrases, clothes fashions, ways of making pots or building arches. Just as genes propagate themselves in a gene pool by leaping body to body via sperm or eggs, so memes propagate themselves in the meme pool by leaping from brain to brain via a process which, in the broad sense, can be called imitation. If a scientist hears, or reads about, a good idea, he passes it on to his colleagues and students. He mentions it in his articles and lectures. If the idea catches on, it can be said to propagate itself, spreading from brain to brain."[7] Of course, it is not necessarily *good* ideas that spread—just *infectious* ones. Think of Cabbage Patch Dolls or Beanie Babies. Could these phenomena exist in a society of rational individuals determining their actions independently of others? Have the afficionados of Gucci jeans come to like them on the basis of independent and thoughtful decisions that they are attractive? Of course not. Advertising is an immense industry for a very good reason: what we think and want can be influenced.

The infectious movement of desires and ideas from mind to mind is even the basis of a new theory of advertising known as "permission marketing." Forget television and radio advertisements, billboards, and the like, where the consumer has to be held hostage for a few moments to receive the message. The new approach is to create infectious desires and ideas that work their way from mind to mind. The front cover of a recent book on the topic tells businesses to "Stop marketing AT people! Turn your ideas into epidemics by helping your customers do the marketing for you."[8] Increasingly, advertisers are banking on the movement of ideas and desires from head to head, this being the very essence of fashion—the emergence of something that generates desire because of its inherent "coolness" and the momentum of an epidemic.

On the other hand, it is fair to wonder if this comparison is anything more than a loose analogy. Do fashions and ideas really spread like viruses and have the potential to erupt into epidemics? To make any kind of judgment, we first need to understand how epidemics work.

THE SECRETS OF SPREADING

IN THE MID-1990S, a severe epidemic of syphilis struck inner-city Baltimore. Before 1993, health authorities in Baltimore were seeing roughly one hundred cases each year. Over the next two years, the number suddenly swelled to nearly four hundred, while the number of newborn infants with the disease shot up by a factor of five.[9] If you plot the number of Baltimore's syphilis cases on a graph according to the year, you will see that the line runs horizontal until 1995, when it suddenly rockets upward. What was going on?

One factor was crack cocaine. In the early 1990s, cities all over the United States were battling a surge in cocaine use, and Baltimore was no exception. Cocaine does not cause syphilus, but its use is associated with the kind of risky sexual behavior that can propel the disease through a population.[10] There were other contributing causes as well. In 1993 and 1994, the city reduced by one-third the number of doctors in Baltimore's clinics for sexually transmitted diseases. Fewer doctors meant that fewer patients were treated with antibiotics. Those afflicted were remaining infectious longer and passing the disease on to a greater number of others.

An ill-timed effort to improve the city's appearance also had an effect. In the early 1990s, authorities demolished a number of old public housing projects and boarded up hundreds of row houses in the downtown area. This region was a center of drug use and prostitution, and as epidemiologist John Potterat suggested, its dislocation had dire consequences: "For years syphilis had been confined to a specific region

of Baltimore, within highly confined sociosexual networks. The housing dislocation process served to move these people to other parts of
Baltimore, and they took their syphilis and other problems with
them."[11]

These were the causes that doctors saw behind the epidemic. None of
them may seem dramatic enough to account for its explosive character.
Incremental changes do not usually give rise to large consequences.
And yet this is the key to the entire argument of *The Tipping Point*. The
epidemiologists and doctors from the Centers for Disease Control in
Atlanta who proposed these ideas are well versed in the ways and means
of infectious diseases. They know that when it comes to epidemics, very
tiny influences can have startling effects.

What determines whether a disease explodes into a nightmarish epidemic or instead quietly fades away? Some diseases move from person
to person more easily than others. All it takes is a cough or sneeze to fill
a doctor's waiting room with flu virus, putting everyone there at risk.
On the other hand, the bacterium *Treponema pallidum* responsible for
syphilis is so fragile that it requires sexual contact to jump between
people. The fate of an epidemic also depends on the population: flu
sweeps through a crowded city far more easily than a sparsely populated countryside, simply because it has more chances to hop.

The proximity of people, in fact, is one of the reasons why we get
colds in the winter. Nevermind what your grandmother always said: we
do not get colds because we are cold. We get them in winter (usually)
because everyone sits around indoors sneezing and spluttering and
wheezing in close quarters, and the virus has a field day.

All manner of complicated details affect how a disease spreads. How
often do people cross paths? How good is the disease at hopping
between people when it has a chance? How long does a person remain
infectious? Once someone has recovered, do they become immune or
can they become infected again? Any estimation of how influenza or
syphilis or HIV, the human immunodeficiency virus, will spread necessarily involves intricate details of biology and the immune response, not
to mention public health measures such as deliberate immunization.
Nevertheless, the situation is not necessarily as complicated as it seems.

Despite the messy details, epidemiologists know that there is a sharp
breaking point between an epidemic and no epidemic, a clear line that
any disease has to cross to make the headlines. The nub of the issue is

this: if one person gets infected, how many others, on average, does this person directly infect?

LITTLE THINGS . . . BIG DIFFERENCES

THE MATHEMATICS IS quite simple. If the number of secondary infections is larger than one, then the number infected will multiply and the epidemic will take off. If it is less than one, the epidemic will die out. There is a third possibility, at least in theory, although it is extremely unlikely. If one infected person, on average, gives rise to *exactly* one further infected person, then the disease will be self-sustaining, but only barely. It lives on the margin of continued existence. Above this point, the disease explodes and a lot of people get infected; below that point, the disease dwindles, eventually disappearing altogether. The balance point is the "tipping point."

In this sense, diseases work much like nuclear reactions. If a pile of uranium is small, there is no "critical mass." The splitting of one atom releases neutrons that can potentially split further atoms. But in a small pile, these neutrons escape from the pile too quickly and so trigger, on average, less than one other splitting. Consequently, a runaway chain reaction never gets going. If there is a critical mass, however, then look out. The neutrons from one split atom will trigger the splitting of more than one atom, and the reaction will feed on itself.

It may seem almost too simple to believe, but even the most sophisticated epidemiological models show precisely the same pattern: a crisp transition between two utterly distinct regimes. In their fancier models, epidemiologists include all kinds of other complications. People are not equally social or sexually active, and so some have a higher chance of passing on a disease than others. You can build these details into a model. Also, for many diseases, an infected person is not always infectious, or may be more infectious at some times than at others. A person infected with HIV, for example, is far more likely to pass on the virus in the first six months. Epidemiologists have intensively studied models with these and many other details over the past few decades, discovering that none of these complications gets rid of the tipping point. There is *always* a tipping point.

This fact explains why scientists of the Centers for Disease Control were pointing to subtle alterations in Baltimore's social and medical reality to explain the syphilis epidemic. All it takes is a few changes to push a disease over the tipping point. Syphilis may have been near the edge already in the early 1990s. One infection may have been triggering, on average, just less than one other infection, and so the disease was keeping itself in check. But then crack cocaine, a few less doctors, and the dislocation of a localized community out into the larger city pushed the disease over the edge—it tipped, and these little factors made a very big difference.

This way of thinking forms the basis of both nuclear reactor safety and much of public health strategy, which are designed to keep nuclear reactions and diseases from going past the tipping point and getting out of control. Of course, for neutrons or for a mindless disease moving through a population, this perspective makes a lot of sense. But does it really have anything to do with Hush Puppies and their stupendous upturn in sales, or with New York City's crime miracle? Can the same picture really apply to the movement of far more nebulous things such as ideas, fashions, and opinions? Where is the science here? In the final analysis, ideas and products and messages are *not* viruses. Perhaps they spread in some totally different way.

As we will see, however, Gladwell is very probably right. The real question in trying to decide about the notion of "idea viruses" is how much the details matter, and the emerging science of networks reveals how to do that. Grabbing hold of the rules for how ideas spread may be difficult. Trying to pin down any mathematical certainties about the process may seem nigh on impossible, but it isn't. Indeed, the theory of networks, and especially the dramatic insight mentioned earlier of the Russian physicist Lev Landau, offers a solid foundation for the ideas of *The Tipping Point*, ideas that, however provocative, might otherwise appear a little vague.

A THEORY OF UNIVERSALS

LEV LANDAU WAS fascinated by the kind of physics that most people, physicists excepted, would find terrifically uninteresting. We all know that when water freezes to make ice, nothing really happens to the water molecules themselves—the change is to be found instead in what those molecules are doing. In ice they sit locked in place like cars in a jammed lot, whereas in a liquid they can move about more freely. Similarly, when gasoline evaporates to vapor or a hot copper wire melts, or when any of a thousand other substances suddenly change from one form to another, the atoms or molecules remain the same. In every case, it is only the overall, collective organization of the atoms or molecules that changes.

Why does a magnetic piece of iron, heated to the temperature of 770°C, suddenly lose its magnetic power? Physicists have known for more than a century that answering a question of this sort means dealing with issues of how one kind of organization can give way to another, and it means learning about what triggers such changes.

In ordinary life, details usually matter. When you write a check, there had better be enough money in the account, and no one lightly ignores the fine print on an insurance policy. In science, details usually matter too. An alteration in just one human gene, after all, can cause cystic fibrosis. Seeking the secrets of sudden phase transitions also means grappling with all kinds of gritty details: the shapes and sizes of the molecules involved, the quantum theory of their electrons, and so on. Even so, Landau argued boldly that most of these details simply do not matter. His aim was to blithely ignore most of the details and to write down one simple theory that would dispense with all conceivable problems of this sort in just one go.

Not surprisingly, Landau's theory took a somewhat abstract form. Landau recognized the influence of a few extremely general forces. To a physicist, for example, heat generally means a violent and disorganized motion of atoms. As a result, higher temperatures tend to work against order—they tend to tear order apart and break it down. On the other hand, lower temperatures let atoms work together more easily and so tend to foster organization. By looking at general organizing and disorganizing tendencies of this sort, Landau worked up a series of equations that described the basic battle between order and disorder and that he hoped would capture the essential magic of any phase transition taking place in any substance whatsover.[12]

This approach is theoretical physics of the very best kind—a big-thinking and ambitious effort to dispel a great mass of confusion in a few steps. And it very nearly worked. As it turns out, Landau was just a tiny bit too bold in ignoring the details. It took three decades of further work by some of the greatest physicists of the twentieth century to work out exactly where Landau had gone wrong, and which details cannot be ignored. The conclusion? There is not just one, unique kind of phase transition. Instead, there are a handful of several different kinds. So Landau was basically right. Even though there are over a hundred elements in the periodic table, and the atoms of each element have their own distinctive properties which go into making the world such a rich and varied place, when it comes to the various ways a collection of atoms or molecules can change form, there are only a few recipes. There is a universal theory of organizational transformation.[13]

How does any of this connect with the tipping point? The modern version of Landau's theory is called the theory of critical phenomena. The word *critical* arises in connection with the peculiar condition that matter gets itself into when poised exactly between two kinds of organization. Water held under just those conditions, for example, is neither a vapor nor a liquid. This is known as the *critical state* since it is the knife's edge between two utterly different conditions. In some very rough sense, it seems a little like the tipping point, but the analogy actually goes much further.

Theorists over the past twenty years have built on Landau's theory, twisting it into a new form that applies not only to changes in organization—between one timeless pattern and another—but also to changes in activity. The new theory, that is, applies to situations in which static, unchanging nothingness transforms into vibrant, changing dynamism, which is exactly what the tipping point is all about. These ideas are at the very forefront of a difficult branch of physics and are not easily described without mathematics. However, it is worth pressing just a bit further, for these ideas may have real implications for the sociology of how influences spread through any culture.

Again, as in the case of Landau's phase transitions, physicists have discovered that details are often totally irrelevant.

MAKING CONTACT

AT THE CENTER of this branch of physics is a little game known as the *contact process*, which works as follows: Imagine a gridwork of people, some "infected" and some not (Figure 19). Think of this as an imaginary world at one instant in time, and let things evolve in the following way. Choose a person at random. If this person is infected, he or she recovers and becomes uninfected. On the other hand, if the person cho-

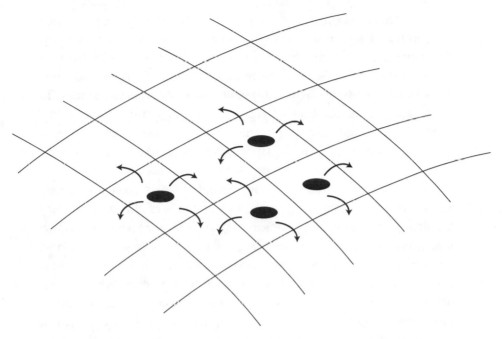

Figure 19. The contact process: a simple model for the spreading of a disease.

sen happens to be uninfected, he or she now has a chance of becoming infected, depending on how many neighbors are already infected. For example, we might set the likelihood of infection to be 10 percent multiplied by the number of sick neighbors. So someone with three sick neighbors would have a 30 percent chance of becoming infected, and so on. Someone with no sick neighbors at all would stay uninfected.

To run this game you just keep choosing people at random and changing their status depending on the situation. The basic idea is that there is a tendency for infection to spread, and simultaneously a tendency for it to disappear. Out of the mix comes a curious transition. Running the game on a computer shows that if the chance of infection is small—10 percent times the number of infected neighbors, for example—then the initial disease will disappear gradually and the grid will be left totally healthy. This is the permanent end of the disease. On the other hand, if the chance of infection is sufficiently high, then the infection will persist and spread, never disappearing. There is a precise threshold for persistence.

The situation obviously bears a certain similarity to the tipping point. There is a regime in which things die out and another in which they do not, a region of inactivity and another of persisting activity. The game as just defined has one specific set of rules. To improve its realism as far as diseases go, you could easily alter the rules and come up with thousands of slightly different and more accurate games. To model the spread of ideas or technological inventions or criminal behavior, you might change the rules in lots of other ways, trying to mimic the essence of those situations. Physicists have studied numerous games of this sort, and this brings us to the punchline.

Given the usual richness of our complex world, you might well expect that altering the rules would have all kinds of effects on the way this game works, revealing thousands of different scenarios for the way inactivity can give rise to activity. But in fact, there seems to be just one. Everything that physicists have discovered indicates that no matter how you bend the rules, there is always a sharp tipping point. What's more, other features are also the same: how far an "infection" tends to spread just above the tipping point, and how quickly it tends to die out just below that point. It is as if the details of how influences spread have no ultimate effect on how things work.[14] Consequently, even though we know very little, perhaps even next to nothing at all about the psychology and sociology of ideas, mathematical physics guarantees that there is a tipping point. All the details that we do not know about are irrelevant to this question.

It is not often that mathematics applies with such power and certainty to social phenomena. The basic idea of the tipping point is not even debatable. No one knows the rules by which ideas spread from mind to mind. No one knows what makes one idea stick and another not, or what the truth is concerning who plays the biggest role in spreading ideas, or what kinds of ideas, behavior, and products are doomed to die on the vine. You can argue about these issues forever. But all these issues have no effect whatsover on the *existence* of a tipping point. However attitudes and beliefs originate and manage to move from head to head, the mere fact that they *can* move, coupled with this striking result from physics, is enough.

So far, however, we have not mentioned the small-worlds idea, or where it fits into this picture. Gladwell suggested that the connectors in a society, those superconnected few, play a disproportionate role in helping influences to spread. This is no doubt true. And yet the real effect of connectors on the spread of influences, and on the nature of the tipping point, is far from obvious. Researchers within the past year have succeeded in wedding the small-world theory with the modern descendents of Landau's theory of universals. And the result adds a delightful new twist to the story, although there are some alarming aspects as well.

BREAKING OUT,
SMALL-WORLD STYLE

AIDS is trying to teach us a lesson. The lesson is that a health problem in
any part of the world can rapidly become a health threat to many or all. A
worldwide "early-warning system" is needed to detect quickly the eruption
of new diseases or the unusual spread of old diseases. Without such a system, operating at a truly global level, we are essentially defenseless, relying
on good luck to protect us.

—Jonathan Mann
Harvard School of Public Health[1]

IN THE TWO centuries preceding the First World War, life expectancy
in Britain soared from seventeen to fifty-two years, mostly because of
better nutrition and hygiene and cleaner air and drinking water. Not so
many children were dying of measles and tuberculosis. In the twentieth
century, the discovery of antibiotics and vaccines cut the death rate
from infectious diseases in the United States by a factor of thousands:
the number of cases of polio, for example, dwindled from 58,000 in
1952 to just 72 in 1965. This was the era of unbridled medical optimism. By 1970, the conquest of all infectious diseases seemed immi-
nent, and the U.S. surgeon general, William H. Stewart, confidently
suggested that it was time to "close the book on infectious diseases."[2]
Sure enough, less than a decade later, health workers totally eradicated
the deadly smallpox virus on a global scale, locking up the last few spec-
imens in laboratories of the U.S. Centers for Disease Control in Atlanta.

In the developed countries today, infectious diseases are not the
major killers they once were, and the public health profession is justifi-
ably proud. But they are also vigilant and cautious, for as we embark on
the twenty-first century, nature is fighting back.

In the developing world, infectious diseases such as tuberculosis, pneumonia, malaria, and measles still take the lives of one in four, killing more than ten million every year.[3] And even in the United States and Europe, biology is inventing new weapons: deadly new viruses and bacteria with countermeasures to our most powerful drugs. The bacterium *Escherichia coli* O157:H7, for example, is a wholly new invention of the biological world, a genetically souped-up version of the older and less troublesome *E. coli* O155. In the past several decades, *E. coli* O155 has engineered resistance to antibiotics. Meanwhile, its mutant progeny, *E. coli* O157:H7, which frequently colonizes uncooked hamburger, has engineered the genes for a far more dangerous toxin, and each year lands hundreds of victims in the hospital.

Far more serious is the spectre of AIDS, the acquired immunodeficiency syndrome. In the twenty years since AIDS first appeared on the radar screens of public health organizations, the epidemic has gone from bad to catastrophic. In 1991, the World Health Organization estimated that by the year 2000 close to eighteen million people worldwide would have the disease. As it turns out, the true number turned out to be thirty-six million, of which more than half had already died. In some countries in sub-Saharan Africa more than 20 percent of the population is now infected. According to a United Nations and World Health Organization report of December 2000, it is "painfully clear that one continent is far more touched by AIDS than any other. Africa is home to 70% of the adults and 80% of the children living with HIV in the world, and has buried three-quarters of the more than 20 million people worldwide who have died of AIDS since the epidemic began. . . . the virus in sub-Saharan Africa threatens to devastate whole communities."[4] In eight African countries, at least 15 percent of all adults are infected. In nations such as Uganda, Rwanda, Tanzania, and Kenya, AIDS will eventually claim the lives of around a third of today's fifteen-year-olds.

There is still no vaccine against the virus that causes AIDS, the human immunodeficiency virus, or HIV. And its ability to mutate rapidly and to present a moving target to researchers raises concerns that there may never be one. Before the AIDS epidemic has run its course, it is certain to have destroyed more lives than the Second World War. Despite centuries of spectacular progress in medicine and biological science, this is not an encouraging moment in the history of mankind.

This era is no longer one of optimism, but one of global disease. As the late Jonathan Mann, professor of epidemiology and immunization at Harvard wrote, "The world has rapidly become much more vulnerable to the eruption and, more critically, to the widespread and even global spread of both new and old infectious diseases.... The dramatic increases in worldwide movement of people, goods and ideas is the driving force.... A person harboring a life-threatening microbe can easily board a jet plane and be on another continent when the symptoms of illness strike. The jet plane itself, and its cargo, can carry insects and infectious agents into new ecologic settings."[5]

This vulnerability is the obvious consequence of the small world and the global network of air travel that stitches it together. As it turns out, however, the social architecture of our world influences the spread of diseases in a number of more subtle ways. Only by facing up to the architecture of the real social world can we appreciate how truly difficult it will be to stop the AIDS epidemic. At the same time, however, a deeper understanding may suggest some powerful clues about the best way to try.

ORIGINS OF AN ADVERSARY

WHERE DID HIV come from? And how did it spread like a poisonous film over the entire globe? Although many details remain contentious, scientists have the basic picture well in hand. As forensic specialists can use DNA evidence to connect a suspect with a bloody shirt discovered at the crime scene, biologists can use DNA to seek the origins of a virus.

In fact, the AIDS virus is not just one virus but two, HIV-1 and HIV-2. And these viruses differ not only in the delicate details of their genetics and structure but also in the diseases they produce. HIV-1 is responsible for the majority of cases in the global epidemic. HIV-2, which has spread mostly to people in western Africa, turns out to be more difficult to catch than HIV-1, and those infected tend to live longer and show milder symptoms. In 1990, researchers worked out the DNA sequences for both of these viruses and then began looking around the biological world for suspects—other viruses with matching DNA fingerprints. Surprisingly, they located two striking candidates in the blood of monkeys living in the jungles of Africa.

HIV-1 is a near-perfect match with a virus called SIVcpz. The letters *SIV* stand for "simian immunodeficiency virus," a virus that infects monkeys. This particular strain of the virus infects chimpanzees, hence, the small *cpz* in the virus's name. In central Africa, most chimpanzees harbor a population of SIVcpz, and yet in contrast to the effects of HIV-1 in humans, SIVcpz is absolutely harmless in chimpanzees. Indeed, it would merit only a tiny footnote in a study of the monkey's physiology were it not for the AIDS epidemic. The virus has been infecting chimps with no ill effects for tens of thousands of years.

HIV-2 has a similar cousin, a simian immunodeficiency virus called SIVsm, which infects another African monkey called the sooty mangabey. An endangered species, this medium-sized, dark-gray monkey survives in the dense jungles of western Africa on a diet of fruits and seeds, and can live to be thirty years old. Again, this virus does no harm to a sooty mangabey. Evidently, mangabeys and SIVsm have been living in a stable relationship for millions of years, and the monkeys have evolved a natural immunity to the infection. The same appears to be true in the case of the chimpanzee and SIVcpz. All this time, however, these viruses represented a catastrophic threat to humanity.

When either of these viruses jumps accidentally to another kind of monkey—a mandrill or a pig-tailed macaque, for example—the infection triggers dreadful AIDS-like symptoms. These other species have not developed immunity and are entirely undefended against an unfamiliar foe. Cross-species infections of this sort have taken place on numerous occasions in zoos and medical laboratories, and also seem to be at the core of the origin of the AIDS epidemic. Scientists believe that sometime in the past, in some remote jungle or village, an SIVsm virus from a sooty mangabey somehow infected a human. It could have been a hunter cutting his finger while making a kill, or someone eating undercooked monkey flesh. On some other occasion, an SIVcpz virus from a chimp made a similar leap. Being less than well adapted to their new human hosts, these viruses would naturally begin evolving rapidly. It is quite reasonable to suppose that SIVcpz and SIVsm could evolve quickly in a few generations into something closely related but ever so slightly different, something much like HIV-1 and HIV-2.

This tale of origins has all the makings of a great mystery. For after the DNA forensics, and the fateful viral transfers to which it points, the story becomes more muddled, not to mention controversial. Tribal Africans have been hunting and eating sooty mangabeys and chimpanzees for tens of thousands of years, and there is little hope of ever identifying dates, places, and people into which the SIV viruses first found a home. This theory of "natural transfer" is indeed natural and plausible, though its details may always remain in the shadows. There is, however, a rival theory.

In *The River*, his magnificent history of the AIDS epidemic, the British journalist Edward Hooper explored the hypothesis that modern medicine itself may have inadvertently introduced the ancestors of HIV-1 and HIV-2 into the human population.[6] His evidence, while circumstantial and ultimately inconclusive, is also not easily dismissed. In 1958 and 1959, polio vaccines were fed to hundreds of thousands of Africans in the former Belgian colonies of the Congo, Rwanda, and Burundi. As it happens, the viruses for these vaccines were grown in the tissues of monkey kidneys. Consequently, if chimps or sooty mangabeys were ever used, there is at least a chance that the vaccines were infected by SIVsm or SIVcpz. Whether this was the case or not, no one can definitely say. Researchers involved in making these vaccines now insist that they always used kidneys from other species of monkeys, in which case any links with AIDS would be ruled out. What really happened? Settling the issue now, fifty years later, may be impossible.

In any event, however the first SIV viruses got into humans, they were clearly there by 1960 at the latest. In 1959, an American doctor named Arno Motulsky traveled to the town of Leopoldville in the Belgian Congo to carry out routine blood sampling in a study of the genetics of the local population, and brought more than 700 five-milliliter samples back to his lab at the University of Washington, where they remained for many years. In the early 1980s, with the AIDS epidemic now raging, André Namhias of Emory University tested the samples and found one to be positive—the earliest AIDS case ever known.

So HIV was spreading in Africa a full twenty-two years before authorities recognized that it was causing a new and very deadly disease. That did not happen until 1981, when five homosexual men presented themselves at hospitals in Los Angeles, gravely ill and suffering from inexplicable symptoms. Why did the epidemic take so long to break out? The HIV viruses may have been around even long before 1959. Since entering into the human population, both HIV-1 and HIV-2 have given rise to numerous additional substrains of the virus. Based on estimates of how rapidly viral mutations accrue, scientists can take a stab at the date when these strains first began diverging, and so try to pin down the moment when their ancestors entered humans. These estimates are not terrifically accurate, but some researchers place the entrance of the AIDS virus into humans as far back as the 1600s, whereas others date it to the 1800s or the early twentieth century.

We face the perplexing question of why the AIDS epidemic only took off in the 1980s. Why didn't it tip earlier? What was it about the complex social network within which the virus was spreading that kept the epidemic in check for many years, and then suddenly let it loose?

BRIDGES BETWEEN WORLDS

ANYTHING THAT HELPS a disease to spread can push it toward the tipping point. Can a virus survive in the open air and go 5 miles on the wind? Can it slip into a hospital's blood supply or the water of a major city? Perhaps most importantly, can it turn "six degrees of separation" to its advantage? We have seen many times how the small-world architecture offers advantages to the brain, the Internet, and the world's ecological communities. When it comes to the spread of a disease, on the other hand, it gives cause for alarm.

In 1968, a flu virus originating in Hong Kong spread in six months to more than fifty countries worldwide, exploiting international air travel routes on its way to killing nearly forty thousand people. In February of 2001, a virus arriving in England, apparently onboard a ship from somewhere in Asia, sparked an epidemic of foot-and-mouth disease in cattle, sheep, and pigs throughout Britain (and Europe, though less seriously). Even before the outbreak had been noticed, the virus had moved from the source farm to seven other nearby farms, then hitched a ride when sheep from one of these farms were trucked to a market. Sheep from this market were subsequently shipped to another market hundreds of miles away. Well before the outbreak was evident, trucks had crisscrossed the country with infected animals.[7]

It is no surprise that long-distance links can help diseases spread quickly. In going from Los Angeles to New York, a virus can clearly benefit from a ride in an airliner or automobile, rather than having to hop monotonously cross-country from person to person and community to community. But the long-distance leaps at the heart of the small world can have far less intuitive effects, and in Africa may have been instrumental in pushing the AIDS epidemic over the edge.

To get some idea of how the architecture of a social network might affect the progress of an epidemic, it helps, as always, to start simple. We will use an extremely crude model of a social network—in this case an "ordered" network—to see how a disease might hop from person to person. Imagine a group of people arranged in a circle (Figure 20), with each "connected" to a few of his or her neighbors. Suppose that everyone is initially uninfected and that some random person then gets an infectious disease. If the infection spreads from person to person along the links, how far will it eventually go? A simple model of this sort is certainly not realistic, but as Argentinian physicist Damián Zanette discovered, it nevertheless offers important insights.

To bring out what is most interesting, however, we first have to make the model a tiny bit more complicated. In its present form, the infection will obviously spread until it sweeps over the entire population. After all, there is nothing to stop it. In reality, infectious diseases do not have it so easy. For most diseases, a few people in the population will be naturally immune. It is also true that being infected does not necessarily make one infectious. Someone infected with syphilus, for example, eventually enters a phase of the disease in which it can no longer be passed on. In the case of flu, most people eventually recover entirely and in any one year, become immune to the strains of flu virus then circulating. Even people with AIDS become effectively noninfectious, as

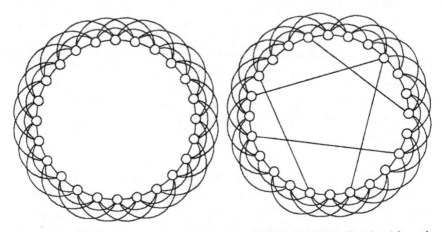

Figure 20. Network evolution. The upper network is fully ordered, with each element being connected to its four nearest neighbors. The lower network comes about through a little rewiring, with a handful of new links thrown in between pairs of elements chosen entirely at random.

most of them actively try to prevent spreading the virus elsewhere—a kind of learned noninfectiousness. In the final analysis, even death itself is a form of immunization.

To include these kinds of features into our network model, we should allow for a third condition. In addition to being infected or uninfected, a person can also become immune, or unable either to catch the disease again or to pass it on. With this one minor complication, the game now works as follows. Everyone in the network starts out uninfected but susceptible to infection. Suddenly, one person becomes infected and the disease begins to spread. As it does, however, some of the infected start to become immune. They may die or recover but in any event become noninfectious. This leads to a situation of immense complexity, for while the disease tries to move from the infected into the uninfected, the noninfectious begin to get in the way. They act as "missing stones" in the network, places where the disease can no longer step while it is trying to spread.

With this complication, it is no longer so obvious how far the infection will go. To find out, Zanette studied a computer model of disease spreading. This approach makes it easy to investigate how the architecture of the social network affects things. You can run the game on a perfectly ordered network like that in Figure 20, or throw in a few long-distance links to fashion a small world. As it turns out, these few links make a world of difference.

MEANS OF ESCAPE

IF YOU GET the flu, you may or may not pass it on to one of your friends, and if you do, which one gets it is largely a matter of chance. Who did you run into at the pharmacy? Who gave you a lift home from work? There is an inherent element of chance in the spreading of disease, and this means that a single infection will not always have the same consequences. Chance necessarily plays a role in our network model too. Nevertheless, with the computer it is easy to run the game thousands of times to see what is and what is not likely, and to see what patterns emerge. The important thing is to learn what is likely to happen.

For the regular network, Zanette found a distinct result: the disease never gets very far. If you start out with one person infected, the disease peters out after infecting only a few others. In fact, as the number of people in the network becomes very large (to mimic any real-world population), the fraction eventually infected in a network of this kind dwindles rapidly toward zero. In one run it may be 1 percent, in a second 1.5 percent, and a third 0.5 percent, but it is always close to zero. So it seems that in a regular network, the effect of the missing stones—the people who have died or otherwise become noninfectious—quickly overwhelms the disease's efforts to spread. As the disease tries to escape from the vicinity of the first infected person, it gets hemmed in by the noninfectious obstacles it has created. In the regular network, the disease never wins the battle.

On the other hand, a handful of long-distance links can tip the balance. If you put a few shortcuts into the network, the fate of the disease becomes increasingly uncertain. With a very few shortcuts, it still dies out quickly, although it begins to make it a bit farther—to infect 2 percent or 3 percent of the population. It is coming closer to breaking out. This gradual change continues until the fraction of long-distance links reaches a critical threshold—about one out of five—when there is a sudden and dramatic transformation. Now the disease either dies out fairly quickly or breaks out to greater things, often spreading into as much as a third of the population. What pushes the epidemic over the edge is not the likelihood of it moving from one person to another, but a change in the very architecture of the social network.[8]

This model may seem remote from any discussion on the origins of the AIDS epidemic. And yet Zanette's small-world "break-out" scenario puts a theoretical basis beneath an explanation based on sociological changes in Africa in the latter part of the twentieth century. In the 1960s, the social organization in western and central Africa changed markedly. People were travelling farther as automobiles and roadways became more numerous, and millions flocked from rural regions into the cities. This movement brought together people from distant villages who otherwise never would have met, effectively establishing a long-distance link between them. Much of this change was triggered by the end of colonial rule in former English, French, and Belgian colonies. Simon Wain-Hobson, a British virologist and AIDS researcher at the Pasteur Institute in Paris, sees the transitional disruption of postcolonial Africa as a potentially key factor in the spread of AIDS: "The English and the French kept people ruthlessly under control. They didn't let them move, they didn't let them travel. And what happened finally when the English pulled out was a free-for-all. There were pogroms, there was corruption, there was blackmail, there were movements of populations, the introduction of the motorcar. There was the beginning of the urbanization of Africa, which is a postwar event."[9]

Health workers also introduced syringes into the region in the 1950s, which were used both in blood sampling and in large-scale vaccine trials. Unfortunately, owing to shortages, syringes sometimes were reused on many people, providing another potential route for the virus to hop from one person to many others.

In 1978, war broke out between Tanzania and Uganda, and the movement of armies acted as another engine for stirring up connections between people who normally lived in distant places. In the autumn of 1978, Ugandan soldiers swept into Tanzania and occupied the "Kagera Salient," a 500-square-mile floodplain of the Kagera River. By April of 1979, the Tanzanian People's Defense Force (TPDF) had reoccupied the Kagera Salient and continued onward, sweeping over all of Uganda and forcing its dictatorial leader Idi Amin to flee to Libya. All the while, the more than forty thousand soldiers of this army, assembled from all over Tanzania, were marching across Uganda and staying, often for weeks or months at a time, in numerous villages.

Scientists generally accept that the Lake Victoria region around the Uganda-Tanzanian border was the first place where the AIDS epidemic really spilled out into big numbers, just after the war ended and through the 1980s. And it seems likely that the movement of Tanzanian troops played a crucial role in helping it to break out. Soldiers are prime targets for AIDS. As a report on AIDS in the armed forces declares, "Military personnel are among the most susceptible populations to HIV. They are generally young and sexually active, are often away from home and governed more by peer pressure than accustomed social taboo. They are imbued with feelings of invincibility and an inclination towards risk-taking, and are always surrounded by ready opportunities for casual sex."[10]

All these changes taken together may have pushed the AIDS epidemic over the edge, replicating something much like the small-world transformation seen in Zanette's network model. As this reveals, it is perfectly possible for a disease to remain effectively sequestered in a small place, infecting a few people but never breaking out, possibly for many years. And it is also possible that a few changes that serve to link together otherwise distant regions can have a dramatic and dangerous effect. It is just possible that AIDS was lurking in a few small villages for more than a century, and that a handful of seemingly insignificant sociological changes, by making central Africa an increasingly small world, provided the catapult that launched the disease out toward every corner of the earth.

When Duncan Watts and Steve Strogatz first built their small-world networks by taking a regular network and adding in a few long-distance links, this was only a first, tentative step into a new geometrical world. The same is true for Zanette's picture of a small-world tipping point for epidemics. As we learned in the last chapter, there is always a tipping point when it comes to the spreading of influences of all kinds. In orthodox epidemiological theory, in fact, the matter seems beyond question, which is comforting for it means that we always have a fighting chance. Indeed, almost everything that public health workers do is aimed at pushing diseases down below their tippings points and keeping them there.

But a closer look at the complex realities of social networks suggests that this strategy may not always be adequate. To tackle some diseases, we may need a more specific "small-world" approach.

NO TIPPING

AS WE HAVE seen in earlier chapters, there are two distinct kinds of small worlds: those with connectors and those without. Connectors, once again, are a superconnected few that possess a disproportionate share of all the links in the network. Sexually transmitted diseases do not move between people by just any route, but spread by navigating on the network of sexual links between people. And as we learned in an earlier chapter, this network appears to be dominated by connectors—it is a small-world network of the aristocratic kind. This pattern is precisely what a study of nearly three thousand randomly selected individuals in Sweden showed: a small world, with a rare few people possessing almost all the sexual links within the community.

How do connectors affect the spread of a disease? We might expect their influence to be akin to long-distance links, making it easier for an infection to move around and pushing the barrier of the tipping point down a little. Long-distance links and connectors alike should lower the bar and make it easier for a disease to tip. But what effects do connectors really have? In recent work, Italian physicist Alessandro Vespignani and Spanish physicist Romualdo Pastor-Satorras discovered that the effects are actually far more dramatic: connectors not only lower the bar of the tipping point but bring it right to the ground. It seems that for diseases trying to spread in an aristocratic network, there simply is no tipping point. Diseases are always tipped and can spread *no matter what.*

This conclusion is weird, but the mathematics bears it out. If a disease cannot hop easily or fast enough, if too many people are vaccinated, or if it kills those it infects too quickly, a single case will generate less than one other new case. As a result, the infection, being below the tipping point, should dwindle and ultimately disappear. What Vespignani and Pastor-Satorras have found is that this is *never* the case in an aristocratic network.[11] Even if a disease has almost zero chance of passing from one person to another—that is, for passing along one particular link—the sheer number of links possessed by the connectors makes it certain that the overall number of infections will grow anyway. The superactive core of the connectors is enough to guarantee that one infection will lead, on average, to more than one. So in a network of this sort, a disease is always "tipped."

This finding is worrying, for since the network of sexual connections in our world is aristocratic, we seem to face a grim possibility that sexually transmitted viruses and bacteria will spread and prevail no matter how low we drive the transmission probability. In this sense, for sexually transmitted diseases at least, the great hope of public health seems to be a fool's chase. As Vespignani and Pastor-Satorras concluded, diseases spreading on aristocratic small-world networks "do not possess an epidemic threshold below which diseases cannot produce a major epidemic outbreak or the onset of an endemic state. [These] networks are therefore prone to the spreading and the persistence of infections, whatever virulence the infective agent might possess."[12]

In the case of HIV, this conclusion offers the truly bleak message that short of a cure or vaccine available to all, the AIDS epidemic will never be stopped. No matter how much money we spend on making it hard for the disease to hop from one person to another, the epidemic will go on smoldering. The human population will live with AIDS indefinitely.

To attack sexually transmitted diseases more effectively, and to have any chance of stopping AIDS, we need some new ideas. Fortunately, knowledge is power. And the network perspective also reveals a few tricks that might offer some hope.

Public health programs ordinarily aim to vaccinate a certain fraction of the population more or less at random, hoping to drive the disease below the tipping point. If there is no such point, this will not work. It might well slow the epidemic, much as thinning the trees in a forest can slow a forest fire, and it might also reduce the size of the epidemic, but it cannot stop it—the continued activity of the connectors will keep it going. To respond adequately, attempts to work against the disease clearly ought to focus on the connectors themselves. This idea may indeed have a chance of working, as several teams of researchers have pointed out that a treatment program targeting connectors can restore the tipping point by altering the very architecture of the social network itself.[13]

For example, suppose it were possible to identify everyone who has had more than, say, twenty-five sexual partners in the last two years. This is one plausible definition of the connectors in the network of sexual connections. If these few people could be effectively immunized, with either drugs or education, then the connectors would, for practical purposes, no longer be part of the network. Consequently, the remaining network would no longer be an aristocratic small-world network and would have a tipping point. In principle, the epidemic could be brought to a grinding halt. Of course, sexual activity being a private matter, and the number of social workers and doctors available to conduct interviews being limited by financial resources, only a fraction of the connectors can ever be identified successfully. Nevertheless, the mathematics reveals that even a fractional success in treating the connectors has a chance of succeeding. The treatment of a very few, but a special few, may be the secret to stamping out the disease.

CORE THINKING

IT IS HARDLY a revelation that the most sexually active people ought to be the targets of public health strategies against sexually transmitted diseases. Epidemiologists for many years have focussed their efforts on what they call core groups, central pools of exceptionally active individuals who keep a disease spreading and propel it further into the community at large. In a series of studies in Colorado Springs in the late 1970s, for example, William Darrow, John Potterat, and other prominent epidemiologists concluded that the great majority of venereal disease cases could be accounted for by a small subset of the population—prostitutes, for one, but also servicemen stationed at a nearby military base and a small cadre of other men and women whose sexual activity was exceptionally high. When it comes to understanding the spread of any sexually transmitted disease, epidemiologists take the core group as absolutely central. And this group is roughly identical to the connectors.[14]

Indeed it is obvious good sense to focus the efforts of health workers on these special few. What is far from obvious is how necessary and truly decisive such focussed programs might be. On the one hand, the connectors in a small-world network of the aristocratic kind are so well connected that their activity dominates all else, and no unfocussed treatment scheme has a prayer of working. Well over 90 percent of all people could be immunized and the disease would keep right on going. On the other hand, treatment of only a fraction of the most highly connected offers a chance to bring the disease under control and possibly wipe it out.

Ironically, when it comes to AIDS, for example, the recipe for stopping the epidemic is not mass treatment and education, but highly selective measures targeted intelligently toward the special few. Taking this insight from complex network theory and putting it into practice will certainly not be easy. But this understanding at least offers epidemiologists and health workers a basic game plan and a strategy that can succeed, not only for the AIDS epidemic but also for new diseases that might arise in the future.

12

LAWS FOR THE LIVING

The triumph of the reductionism of the Greeks is a pyrrhic victory: We have succeeded in reducing all of ordinary physical behaviour to a simple, correct Theory of Everything only to discover that it has revealed exactly nothing about many things of great importance.

—*Robert Laughlin and David Pines*[1]

*R*EDUCTIONISM, ACCORDING TO *The Concise Oxford Dictionary*, is defined as "analysis of complex things into simple constituents; the view that a system can be fully understood in terms of its isolated parts, or an idea in terms of simple concepts."[2] If your car does not work, the mechanic looks for a problem in one of the parts—a dead battery, a broken fan belt, or a damaged fuel pump. Similarly, pain in a patient's leg alerts a doctor to a broken bone or bruised muscle, and an engineer trying to fix a computer looks for a burnt-out microchip or a software glitch. Reductionism is the idea that the best way to understand any complicated thing is to investigate the nature and workings of each of its parts. This approach is how we solve problems, and it comprises the very basis of science.

Nevertheless, it is not always true that a system, as the definition above would have it, can be "fully understood in terms of its isolated parts." As we have seen, no understanding of viral or human biology by itself offers a clue about the network effects that strongly influence the AIDS epidemic. Similarly, the stability of the global ecosystem cannot be ascertained merely by studying the biology of creatures in isolation. Grasping the nature of parts in isolation offers little hint of how they might work in combination. This isn't to say that reductionism is a bad idea—only that the definition just given is inadequate.

A better definition would assert that a system can be fully understood in terms of its parts *and the interactions between them*. And yet it is also important to acknowledge that the interactions between the parts of a complex network often lead to global patterns of organization that cannot be traced to the particular parts. Network architecture is a property not of parts but of the whole, as is the existence or nonexistence of a tipping point. In social networks, global effects of this kind can have surprising and important consequences, as a sociologist from Harvard named Thomas Schelling pointed out over thirty years ago. What is the origin of racial segregation, for example? In the United States, the persistence of segregation is usually attributed to racism or biased practices on the part of the government or the real estate industry. But another less obvious factor may be equally influential.

Schelling began by imagining a society in which most people truly desire to live in balanced and racially integrated communities, with just one minor stipulation: most people would prefer not to end up living in a neighborhood in which they would be in the *extreme* minority. A white man might have black friends and colleagues and might be happy to live in a predominantly black neighborhood. Just the same, he might prefer not to be one of the *only* white people living there. This attitude is hardly racist and may indeed be one that many people—black, white, Hispanic, Chinese, or what have you—share. People naturally enjoy living among others with similar tastes, backgrounds, and values.

Nevertheless, innocent individual preferences of this sort can have startling effects, as Schelling discovered by drawing a grid of squares on a piece of paper and playing an illuminating game. On his grid, he first placed at random an equal number of black and white pieces, to depict an integrated society of two races mingling uniformly. He then supposed that every piece would prefer not to live in a minority of less than, say, 30 percent. So, taking one piece at a time, Schelling checked to see if less than 30 percent of its neighbors were of the same color, and if this was the case, he let that piece migrate to the nearest open square. He then repeated this procedure over and over until finally no piece lived in a local minority of less than 30 percent. To his surprise, Schelling discovered that at this point the black and white pieces not only had become less uniformly mixed but also had come to live in entirely distinct enclaves.[3] In other words, the slight preference of the individual to avoid an extreme minority has the para-

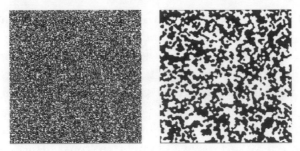

Figure 21. The inexorable results of Schelling's segregation model. At left, the population is initially well mixed and the community is highly integrated. But as people move to avoid being in areas where they are in the extreme minority, the population naturally segregates into clusters of nearly uniform black or white. (Images courtesy of Nigel Gilbert and Klaus Troitzsch, reprinted by permission.)

doxical but inexorable effect of obliterating mixed communities altogether (Figure 21).

This finding is not meant to prove that racism or unfair institutions have nothing to do with perpetuating ethnic segregation. But it does reveal how surprising kinds of organization can well up quite naturally in complex networks, and how imperative it is to look beyond "parts in isolation" in trying to comprehend such effects. Even if every trace of racism were to vanish tomorrow, there may still be a natural tendency for races to separate, much like oil and water. Social realities are fashioned not only by the desires of people but also by the action of blind and more or less mechanical forces—in this case forces that can amplify slight and seemingly harmless personal preferences into dramatic and troubling consequences.

As we have seen earlier, the small-world theory offers a resolution of the social mystery of six degrees of separation. But this is only a part of network theory, a part that is more generally about discovering broad patterns of organization that well up naturally in networks of all kinds. Today, a growing number of researchers are following Schelling's lead and exploring such network effects in a variety of social settings. Whether this approach reaches "beyond reductionism" or not isn't important. What is important is that this more sophisticated perspective offers new hope for the social sciences and, in particular, for economics.

THE DEMISE OF *HOMO ECONOMICUS*

FREE-MARKET THEORIES of economics date back at least to the Scotsman Adam Smith in the latter half of the eighteenth century. In his famous *The Wealth of Nations*, Smith claimed that free trade among the members of a society, each intending only his own selfish gain, would nevertheless lead to an outcome that would be good for the society as a whole. After all, as he noted, "It is not from the benevolence of the butcher, the brewer, or the baker that we expect our dinner, but from their regard to their own interest. We address ourselves, not to their humanity, but to their self-love, and never talk to them of our own necessities but of their advantages."[4] According to Smith, a society of individuals, each rationally pursuing their own advantages, offers the best way to achieve the public good. If an individual can profit by manufacturing some product or by supplying some service, he will do so. And his very ability to do so proves that other members of the society must desire those goods or services. As a natural result, the full spectrum of society's needs will be met, even though individual gain is the engine of its success. A free-market economy of this sort should work both smoothly and efficiently, even without any global management, as if guided and organized, in Smith's well-travelled metaphor, by an "invisible hand."

Today, Smith's metaphor stands at the very center of the entirety of Western economic thinking. And for more than a century, an army of theoretical economists has worked diligently to prove that it is indeed true—that individual greed really must translate into collective good. To do so, they generally assume that economic agents are not only greedy but also perfectly and infallibly rational. No one, the theory demands, would be so daft as to let their emotions get in the way of rational decision making, or would simply imitate others without excellent reasons for doing so. As one economist describes the theory, "People are assumed to want to get as much for themselves as possible, and are assumed to be quite clever in figuring out how best to accomplish this aim. Indeed, an economist who spends a year finding a new solution to a nagging problem—such as the optimal way to search for a job when unemployed—is quite content to assume that the unemployed have already solved the problem and search accordingly."[5] This is eminently silly, of course, and yet there is a method here. With the slate of ignorance regarding human behavior wiped clean, theory can proceed as if it were dealing with atoms or molecules, and can "prove" that the invisible hand really works.[6]

Almost every corner of contemporary economic theory remains obsessively fixated on this "cult of rationality," whether it is applied to explain the structure of the business firm or the distribution of wealth. Fortunately, this inflexible theoretical tradition is slowly giving way to a more mature perspective—one that is willing to deal with the complexities of economic reality, rather than pushing them into the background in order to simplify the mathematics. A growing number of researchers have embraced the new field of "behavioral economics," which aims to address the shortcomings of traditional theories by accepting human irrationality and trying to found economic theories on a more realistic picture of human behavior.[7] In this way, for example, researchers have built strikingly realistic models for financial markets in which stock prices fluctuate with great irregularity as investors' tendency to imitate one another leads to crashes and bubbles based on herd behavior.[8]

But researchers are also working in another direction—by recognizing that in networks of interacting things, even if those things are people, the details sometimes do not matter. Whether people are rational or irrational or something else altogether, the details of their behavior may have little effect on some of the most basic of all economic realities. And there are many notable patterns in the economic world, some of which have been known for more than a century. In every country, for example, there are very few rich and many poor. You might well expect the balance between the two to be different in each country, as each has its own particular products and skills, some thriving on agriculture or heavy industry, others on high technology and so on. Given the world's many distinct cultures and their histories, there would be little reason to expect any general patterns in the way wealth accumulates. But in 1897, an Italian engineer-turned-economist named Vilfredo Pareto found otherwise. Pareto discovered a pattern in the distribution of wealth that appears to be every bit as universal as the laws of thermodynamics or chemistry.

AN ECONOMIC UNIVERSAL

SUPPOSE THAT IN Germany or Japan or the United States you count up how many people have, say, $10,000. Next, repeat the count for many other values of wealth, both large and small, and finally plot your results on a graph. You would find, as Pareto did, that toward the end of the graph corresponding to small wealth, there are many people, and that as you progress along the graph toward large wealth, the number of people decreases. But Pareto studied the numbers more closely and discovered that they dwindled in a very special way: toward the wealthy end of the distribution, each time the value for wealth is doubled, the number of people who have that much falls off by a constant factor.[9] Up-to-date numbers show the same pattern for countries all over the earth. In Japan, for example, the constant factor turns out to be close to four.[10]

This pattern represents another of those fat-tailed distributions we have mentioned many times. As discussed earlier, if you measure the heights of a thousand people, they will fall into a narrow pattern about some well-defined average. The resulting bell curve implies that great deviations from the average are very rare. The distribution of wealth decidedly does not work the same way. The curve of Pareto's law falls off much more slowly than the bell curve, implying that there is a significant number of extremely wealthy people. Sometimes referred to as the *80-20 principle*, this pattern implies that most of the wealth gathers in the pockets of a few. In the United States, for example, approximately 20 percent of the people own 80 percent of the wealth. In Mexico, Argentina, Britain, and Russia, or in the nations of Western Europe, the numbers are similar. But this 80-20 distribution is not really the point; in some cases, the precise numbers might be 90-20 or 95-20 or something else. The important point is that the distribution—at the wealthy end, at least—always follows a strikingly simple mathematical curve. And, as a result, a small fraction of people always own a large fraction of the wealth.

What kind of regularity in human behavior or culture could lead to this pattern? Is there some devilish conspiracy among the rich? Not surprisingly, given the strong emotions stirred by matters of wealth and its inherent disparity, economists have flocked to the question. Of the central issues in economics, as John Kenneth Galbraith wrote, the first is "how equitable or inequitable is the income distribution. The explanation and rationalization of the resulting inequality has commanded some of the greatest, or in any case some of the most ingenious, talent in the economics profession."[11] From a mathematical point of view, however, Pareto's law has stubbornly defied explanation. Of course, it might simply be that a few people are exceptionally more creative and talented than most, and so an explanation of the wealth distribution would require first an explanation of the inherent spread in human talents and abilities—a tall order indeed. On the other hand, there might be another possibility.

How might one even go about trying to explain Pareto's law? When it comes to explaining why one individual is rich and another is not, there is no substitute for the historical method. One has to delve into the details of inheritance and education, inherent money-making ability and desire, as well as plain old luck. The sons or daughters of doctors or bankers frequently become doctors or bankers themselves, whereas children born into inner-city poverty often remain mired in poverty, unable to escape their environment. Understanding why Bill Gates, cofounder and chief executive officer of Microsoft, is so wealthy is much like understanding why the great Mississippi River enters the Gulf of Mexico near New Orleans rather than somewhere else. Gates's particular background, education, and temperament, combined with the possibilities and opportunities of the computer revolution, fit together in a unique historical way to bring him enormous wealth.

But Pareto's law is not about individuals. Rather, it captures a pattern that emerges at the level of many individuals and leaves individual histories aside. In this sense, the pattern discovered by Pareto is like the global organization pattern for river networks noted in chapter 6, which refers not to the shape of this or that stream but to the overall network structure of the river basin. In that case, scientists have discovered that order emerges in the collective network even though accidents rule the day at the level of individual streams. In principle, the same might be true of wealth. Perhaps Pareto's law reflects little about human cultural, behavioral, or intellectual features and arises instead as a consequence of some deeper principle of organization.

WEBS OF WEALTH

TO EXPLORE THE distribution of wealth, let's forget for the moment about creativity and risk taking, the distribution of intelligence, and other factors that might come to mind, and instead focus on the raw essentials. An economy, all would agree, is a network of interacting people. Some may act as individuals, and others may act within larger or smaller organizations, but we can leave these complications aside for now. To make things visual, imagine a large number of people in a network. The network might be a random network, an ordered network, a small-world network, or what have you. Whatever type of network it is, each person has a certain wealth, and over the days and weeks, this wealth will change—usually in one of two fundamental ways. To begin with, people buy and sell products and services of all kinds. When your monthly paycheck arrives or when you sell your car, your financial wealth goes up. If you go on vacation in Italy or build a new patio, it falls. Such transactions shift wealth from one person to another. But wealth can also be created or destroyed. If you purchase a house or a piece of land, its value may rise or fall. Similarly, everyone who invests in the stock market gambles on its movements. During the 1990s the market soared, creating vast quantities of totally new financial wealth.

The point is that a person's wealth can go up or down either by transactions with others or by earning returns (positive or negative) on investments. This is hardly news, but it implies that in the web of wealth, two factors should drive the numbers up and down. As people get paid, pay rent, buy food, and so on, wealth should flow through the network in a more or less regular way, like water through pipes. Meanwhile, owing to investments, each person's wealth should show a slow upward trend with additional random kicks up or down as their investments do better or worse.

Obviously, this picture leaves out almost every detail of reality except the very most basic. And yet it is intriguing to wonder if even these basic factors might imply something about how wealth will end up being distributed. A couple of years ago, physicists Jean Philippe Bouchaud and Marc Mézard of the University of Paris took a large step toward answering this question by bringing into the picture one further "obvious" fact—that the value of wealth is relative. A multimillionaire, for example, will not ordinarily sweat losing a few thousand dollars on the stock market. But the same loss would likely be catastrophic for a single parent trying to raise her son while putting herself through college. The value of money depends on how much one already has, and consequently, wealthy people tend to invest more than the less wealthy do.

With this simple observation, Bouchaud and Mézard found that they could turn the network picture into a set of explicit and fundamental equations to follow wealth as it shifts from person to person, and as each person receives random gains or losses from their investments. With equations in hand for a network of 1,000 people, the two physicists set to work with the computer to see what they might imply. Not knowing precisely how to link people together into a network of transactions, they tried various patterns. And unsure of how precisely to set the balance between the importance of interpersonal transactions versus investment returns, they tried shifting the balance first one way and then the other. What they discovered is that none of these details alters the basic shape of the wealth distribution.

Giving people random amounts of wealth to start out, and letting the economy run for a long time, Bouchaud and Mézard found that a small fraction of the people always ended up possessing a large fraction of the entire wealth. What's more, the precise mathematical distribution followed Pareto's law *exactly*—in excellent correspondence with data from the real world.[12] This result occurred despite the fact that every person in the model was endowed with identical "money-making" skills, suggesting that differences in talent may have little to do with the basic inequality in the distribution of wealth seen in most societies. Rather, what appears is akin to a fundamental law of economic life, a law that emerges naturally as an organizational feature of the network.

This may seem like getting something for nothing, and in some ways it is. But the discovery suggests that the temptation to find complex explanations behind the distribution of wealth may be seriously misguided. The model has more to teach, however.

SHADES OF INEQUALITY

WHAT MAKES WEALTH fall into the pockets of a few? The secret appears to be quite simple. On the one hand, transactions between people tend to spread wealth around. If one person becomes terrifically wealthy, he or she may start businesses, build houses, and consume more products. In each case wealth will tend to flow out to others in the network. Likewise, if one person becomes terrifically poor, less wealth will flow out, as that person will tend to purchase fewer products. Overall, the flow of funds along links in the network should act to wash away wealth differences.

As Bouchaud and Mézard discovered, however, this washing-out effect never manages to gain the upper hand, for the random returns on investment drive a kind of "rich-get-richer" phenomenon that is not easily defeated. Out of 1,000 people, no two will share exactly the same investment luck. Most will win about half the time and lose about half the time, while a few will be wildly lucky and others will meet with disaster. But recall that those with more wealth tend to invest more and so have a chance to make greater gains still. Hence, a string of positive returns builds a person's wealth not merely by addition but by a process of multiplication, with each subsequent gain growing bigger. This is enough, it turns out, even in a world of equals, all of whom receive entirely random investment returns, to stir up huge wealth disparities in the population.

Nevertheless, Bouchaud and Mézard's network model reveals that there are degrees of inequality and that Pareto's law can be influenced. Recall that in the fat-tailed pattern, each doubling of wealth leads to a decrease by a constant factor in the number of people having that much. The factor might be 1.8 or 2 or 3.4 or another number. Wealth in every case is always owned disproportionately by a small fraction of the most wealthy, but as the number in the fat-tailed pattern grows smaller, the concentration of wealth becomes more pronounced. At one value, 10 percent of the wealthiest may own 90 percent of the wealth, while at another 5 percent may own 98 percent. In other words, the fat-tailed pattern can remain, even while the precise numbers change.

In this regard, the model offers a general message: encouraging exchange between people, with other things being equal, will tend to distribute wealth more equitably. Bouchaud and Mézard found greater equality whenever they boosted the flow of wealth along the links or increased the number of such links. Alternatively, stirring up the wildness and unpredictability of investment returns worked in the opposite direction, which is not surprising, as it boosts the influence of the rich-get-richer phenomenon. Of course, this model is so abstract that it is not meant to provide detailed recommendations for public policy. Nevertheless, it may offer some very basic recommendations, some obvious and some not so obvious, about how the distribution of wealth might be altered.

The model reveals, for example—and this should come as no surprise—that taxation will tend to erode differences in wealth as long as the money is redistributed to the society in a more or less equal way. After all, taxation corresponds to the artificial addition of some extra links into the network, along which wealth can flow from the rich toward the poor. Taxation does not alter Pareto's law, but the wealth will become distributed somewhat more equitably, with the rich owning a smaller fraction of the overall pie. Somewhat more surprisingly, the model suggests that a like redistribution of wealth should result from any economic measures aimed at boosting spending right across the economy. Broad taxes on the sales of luxury items, for example, may even tend to increase wealth disparities.

The model might well also offer an excellent testing ground for some of the arguments that politicians often use to justify various policies. In the United States, for example, the 1980s and 1990s were dominated by free-market ideology and government deregulation, much of it defended by the idea that wealth would "trickle down" to the poor. Everything was done to encourage investment activity of all kinds, regardless of high risks, possible environmental damage, and so on. It was no accident that this was the era of junk bonds and the savings and loan debacle. Did the wealth trickle down? Based on the network picture, there is little reason to expect that it would. In fact, one would expect just the opposite. A dramatic increase in investment activity, unmatched by measures to boost the flow of funds between people, ought to kick up an increase in wealth inequality, which is indeed what happened. Today the United States has a significantly less equitable distribution than it did three decades ago. Wealth is more highly concentrated in this country than it is in European nations, and is verging on the level seen in Latin American countries.

To say it again, however, the point of this network model is not to offer principles for guiding economies with the same precision that NASA guides rockets to the moon. This model is meant to be a starting point. Even though its conclusions are highly abstract, at least they are not based on dubious assumptions about human psychology, perfect rationality, and the like, or on vague and unverifiable speculations on what might or might not trickle where. By trying to get by with as few assumptions as possible, and thereby achieving limited yet believable results, the network perspective offers an encouraging beginning in this direction.

To be able to make intelligent and well-informed social decisions, we need to have a deeper picture of where important economic patterns such as the distribution of wealth come from, and what basic forces affect them. Otherwise, we might be in for some disturbing surprises.

In this regard, there is another truly alarming economic message buried within this web of wealth.

A PROBLEM OF CONDENSATION

AS WE HAVE seen, the irregularity of investment returns stirs up wealth differences, while transactions of all types between people tend to wipe them out. The competition between these two forces leads to Pareto's law, with a greater or lesser concentration of wealth falling into the hands of a small fraction of people. In studying their model, however, Bouchaud and Mézard discovered that if the investment irregularities grow sufficiently strong, they can completely overwhelm the natural diffusion of wealth provided by transactions. In this case, an economy can pass through a sudden and dramatic transition in which the wealth disparities kicked up are simply too pronounced to be adequately tempered by flows between people. The economy will tip—and wealth, instead of being possessed merely by a small minority, will instead "condense" into the pockets of a mere handful of super-rich "robber barons."

This may not sound dramatic, but in a society of millions, it would be. The wealthiest 10 percent of the U.S. population is a group of 300,000 people, and the collapse of their wealth into the hands of five or six people would represent a dramatic transformation of society. An attendant power shift would follow, with potentially great political ramifications. As worrying as this scenario might be, it is not science fiction. Even though the network model is abstract, this property is also its advantage, for it proves on fundamental mathematical grounds, with few disputable assumptions, that a tipping point of this sort must exist in any economic society. The economy in the United States, for example, may currently be far from this point, or close to it. No one knows. In any case, policy makers ought at least be aware of the precipice over which an economy might tumble.

It is intriguing to wonder whether some countries, particularly developing nations, may already be in this "condensed phase." It has been estimated, for example, that the richest forty people in Mexico have nearly 30 percent of the money. It could be, also, that economic societies may have existed in this phase quite frequently earlier in history. Long-term economic trends this century lend some credence to this idea, as the total share of the richest in England, for example, has fallen over the last century, particularly in the period 1950–1980.

On the other hand, political instability may also offer the opportunity for an economy to plunge into this phase. In Russia, following the collapse of the USSR, wealth has become spectacularly concentrated, with inequality dramatically higher than that in the West. No one can be sure why, but the model would suggest that both increased instability and lack of opportunities for wealth redistribution might be at work. In the social vacuum created by the end of the Soviet era, there are few regulations to protect the environment or to provide safety for workers, and so economic activity is less restricted than in the West. This lack of restriction not only leads to pollution and untold human exploitation but also means huge profits for some companies while others fail entirely. Economists have also pointed out that Russia has been slow to implement taxation measures that would help to redistribute wealth.[13] Taxation is a form of enforced trade, and without it, wealth discrepancies rapidly escalate.

Again, this simple model is not the end in explaining the distribution of wealth or in offering guidance concerning how best to manage it. Nevertheless, by starting with remarkably simple assumptions and studying the patterns that emerge inevitably in a network of interacting agents, Bouchaud and Mézard have succeeded in explaining one of the most basic patterns ever observed in economic life. Their work reflects just one example, however, of a new kind of economics now emerging in research centers around the world. Given our limited understanding of human behavior, is it even possible to build accurate economic theories? The answer clearly is a resounding yes. To be sure, few details in economic life may be predictable. It may be just as impossible to tell which particular individual will end up wealthy or which particular firm will be a success as to foretell the movements of the stock market. But the patterns of economic law arise at the level of many people or many firms, or in the statistics of price fluctuations over the longer run. As it turns out, both price fluctuations on the market and the size of business firms follow fat-tail patterns very similar to Pareto's law and can be explained with similar simplicity. Ignorance at the level of human behavior does not necessarily dictate ignorance at the level of collective economic functioning.

13

BEYOND COINCIDENCE

Year after year we are becoming better equipped to accomplish the things
we are striving for. But what are we striving for?
—*Bertrand de Jouvenel*[1]

THINGS ARE OFTEN simpler than they appear. And coincidences are
not always what they seem. If you are snorkeling in the blue waters off
Bermuda and bump heads with a friend of your dentist's sister, this is
not a coincidence. Indeed, many events during that day could have
gone differently. That specific encounter was indeed coincidental—an
accident from which no lessons can be drawn. Behind such accidents,
however, lies an identifiable engine, for it is not only a few people who
are only a few links away—almost everyone is. We do not notice the
small-world encounters that almost take place but narrowly miss for
one reason or another. On the crowded street of any major city, there
can be no more than a few links separating any two individuals—we are
all that close. What is surprising, in fact, is how infrequently we notice
the small world, and how often we believe that most others are indeed a
long way away.

On another level, it is also not a coincidence that the wiring of the
human brain turns out to have the very same small-world structure as
our social networks, nor that these patterns turn up again in the Internet and the World Wide Web, in the way words link together in human
language, or in the food webs that underlie the world's ecosystems. If
this were indeed merely a coincidence, it would be a truly spectacular
one. But things *are* simpler than they appear. If social networks grow
through accident, influenced by cultural and economic events; if neural

connections in the brain follow from the demands of efficiency as enforced by evolution; if the Internet has been stitched together happenstance, while meeting the needs of commerce and technology, common threads run through the world and link these diverse networks together.

As Duncan Watts and Steven Strogatz discovered, a few long-distance links thrown into an otherwise gridlike network will suffice to make a small world. As Albert Barabási and Réka Albert noticed, the simplest of all conceivable patterns of growth—the richest and most popular getting still richer and more popular—leads to small-world networks of a slightly different kind. From two very simple rules follow small worlds of many kinds—this is no coincidence.

If the project of history is to tell stories that explain differences, science is largely about discovering and exploring similarities. Or as Herbert Simon put it so well, "To find meaningful simplicity in the midst of disorderly complexity."[2] Whether we discover a similarity in discussion with a new acquaintance, or instead in the laboratory, it is always a surprise—we suddenly link disparate experiences and stumble over deeper principles. In a world that could conceivably be random and lacking in any discoverable order, scientists have discovered instead that order abounds, even within the context of overwhelming disorder.

The very aim of the science of complexity is to discover patterns in complex networks of all kinds and to learn how we might use this understanding to better ourselves and our world. Central to this task is the notion of *emergence*, the idea that meaningful order can emerge all on its own in complex systems made of many interacting parts. In the economic world, Adam Smith's invisible hand and Pareto's law of wealth distribution represent two principle emergent properties. Of course, recognizing these patterns and understanding their origins is only one step; we also want to know how we might influence them and how to use network properties to our advantage.

Before finishing this book, I'd like to explore very briefly some of the lessons to be learned from the network discoveries explored here. For example, what does the small-world theory counsel as far as building efficient organizations, or communities that work well? It is certainly true that we are far closer to the beginning of network science than to its ending, if there will ever be one. Even so, a number of practical lessons may already be evident.

CAPITALIZING ON THE SMALL WORLD

WHEN IT COMES to network architecture, the small-world network offers obvious advantages because of its intimacy. For a computer network or a nervous system, or for a company of people who need to organize their efforts, this pattern of connectivity fosters rapid communication between disparate elements—computers, neurons, or employees. Recall, however, that random networks also have only a few degrees of separation. What distinguishes a small-world network is not only that it has a low number of degrees of separation but also that it remains highly clustered. We might say that the fabric of the network is densely weaved, so that any element remains comfortably and tightly enmeshed within a local web of connections. Consequently, the network overall can be viewed as a collection of clusters, within which the elements are intimately linked, as in a group of friends. A few "weak" links between clusters serve to keep the whole world small.

If the long-distance links keep the network well connected, clustering offers numerous strong bonds and a context within which each element is firmly embedded. When it comes to economic life, it is easy to imagine how long-distance links might be advantageous, to make transactions more rapid, for example. But clustering has an equally crucial effect, if one that is slightly more subtle. Over the past few decades, a growing number of economists have come to acknowledge that pure and perfect rationality cannot fully explain the behavior of individuals. Our rational powers are obviously more limited. Still, most economists insist that individuals almost always try their best to realize their own benefit, and that even if they are not fully rational, people are greedy and generally act so as to maximize their own gain. As Mark Granovetter has argued, however, even this more limited perspective leaves out central features of real economic life.[3]

Granovetter points out that in any organization or in a family or group of friends, relationships established over a time can also be the source of behavior with economic consequences. Only rarely do individuals act as isolated beings, able to pointedly pursue their own personal agenda; more frequently, we act within the context of numerous other goals and constraints originating in our social life. And many of these goals and constraints have little to do with economic ends, but more to do with conforming to a set of shared norms and ethical values. As Granovetter made the point more recently, "Any account of human interaction which limits explanation to individual interests abstracts away from fundamental aspects of relationships which characterize economic as well as any other action. In particular, horizontal relationships may involve trust and cooperation, and vertical relationships power and compliance, well beyond what individuals' incentives can explain. Trust and power drive a wedge between interests and action."[4] For example, we obey the commands of a superior, forego some opportunity because it would not seem fair to a friend, and so on. Human behavior is not merely economics, and relationships and the ethical values on which they rest are as important as individual goal-seeking behavior.

What are the effects of these social bonds? They may, of course, be many. Over thirty years ago, for example, Stanley Milgram pointed to the consequences of such social ties when trying to explain the disconcerting results of his "obedience to authority" experiments. As individuals, most of us would be quite unwilling to torture an innocent person. We are socialized from birth to see this as unacceptable. And yet the volunteers in Milgram's experiment were not isolated individuals. By agreeing to work with the experimenter, each had already stepped into a social network, and the consequences of this fateful step were greater than they might have imagined.

The social world gains much of its efficacy from our unquestioning acceptance of authoritative relationships. No military unit could work effectively if each man or woman acted solely on his or her own account—indeed, much of military training consists in bringing people to accept themselves as cogs in a greater whole, a whole that can be coordinated from above through a hierarchical common structure. Authoritative relationships play similarly central roles within families, businesses, governments, schools, and so on, where clearly delineated and accepted lines of authority promote efficiency and keep interpersonal friction to a minimum. There is, however, a price, for an individual, in accepting a position within such a network, necessarily cedes a certain amount of individual control and suppresses personal autonomy so the group can function.

Consequently, as Milgram suggested, the startling degree to which volunteers in his experiments would follow commands may simply reflect the fact that humans have been socialized to live within groups.

As Milgram made the point in reference to the experiment, "The person entering an authority system no longer views himself as acting out of his own purposes but rather comes to see himself as an agent for executing the wishes of another person. Once an individual conceives his action in this light, profound alterations occur in his behavior and his internal functioning. These are so pronounced that one may say that this altered attitude places the individual in a different state from the one he was in prior to integration into the hierarchy."[5]

Of course, the consequences of such "social embeddedness"—a network effect if there ever was one—are by no means always or even frequently negative. Another, explored recently by the American political scientist Francis Fukayama, is the production of so-called social capital, a term first used by the sociologist James Coleman which refers to a determinant of economic success that traditional economics leaves wholly out of the picture. Fukayama suggested that the ability of a nation, a community, or a company to compete economically is instrumentally influenced by the inherent level of trust among its members. Trust represents an intangible form of social capital, which Fukayama defines as "a capability that arises from the prevalence of trust in a society or in certain parts of it. It can be embodied in the smallest and most basic social group, the family, as well as the largest of all groups, the nation, and in all the other groups in between. Social capital differs from other forms of human capital insofar as it is usually created and transmitted through cultural mechanisms like religion, tradition or historical habit."[6] Loosely speaking, social capital is the ability of people to work together easily and efficiently based on trust, familiarity, and understanding. Its importance lies in its power to create efficient networks for transactions. For example, in a network of companies that involve people linked through networks endowed with social capital, the costs associated with forging legally binding contracts are reduced. Decisions are made more readily and quickly because of shared norms and goals, and so on.

Curiously, social capital appears to be intimately connected with the mathematical property of clustering that we have met in exploring network architecture.

SOCIAL WHOLENESS

IN A CLUSTERED network, most of the links between people are strong links, endowed with history and cemented with frequent interaction. What's more, a large fraction of people within the network share such links. This is true whether we speak of a network of friends, colleagues in an office, or soldiers in a unit—shared experience and proximity through time build ethical feeling and shared norms. In an organization, for example, a new employee learns the ropes by imitating the behavioral patterns of the other employees with whom he or she interacts. Perhaps most of what is learned is not by explicit instruction but by nonverbal communication as the new employee comes to see how colleagues behave and how the organization works.

Generally, the sharing of norms and expectations that results furthers the organization's goals. After all, the members of the group share a great many tacit principles of behavior, which makes innumerable transactions more efficient. Hence, the mere fact of clustering in a social network tends to foster social capital that would not otherwise arise. What are the consequences? If Fukayama is right, then they are appreciable and may often spell the difference between economic success and failure. A few examples may serve to illustrate the point.

Fukayama contrasted, for instance, the organization and behavior of workers in German and French factories. In Germany, a shop foreman on the floor will often know how to perform the jobs of most or all of the people working under him. If necessary, he has no qualms getting in there and taking up that job himself. The foreman is not merely an overseer but an intimate and trusted component of the group. He evaluates workers on the basis of his own experience and observations of their work, and he can move workers from one job to another to improve efficiency.

In a French factory, in contrast, the shop foreman's relations with his workers tend to be both more formal and less effective, as they are based less on an established form of trust. For historical and cultural reasons, the French tend not to trust superiors to make honest personal evaluations of their work. And numerous rules set out by a ministry in Paris stipulate what the foreman can and cannot do. He cannot move workers from one job to another, and so efficiency necessarily tends to suffer. The lack of bonds of trust destroys or inhibits the sense of team-

work, and offers resistance to the introduction of new and more efficient manufacturing schemes.

In many ways, social capital is the ability of a team to work as a team
on its own, willingly, without participation being managed by legally
binding rules and regulations, the need for which is already a signal of
lacking efficiency. In the workplace, efficiency and the organization's
ability to learn and adapt in the marketplace tend to suffer from a lack
of social capital. The same lack can hamper efforts to improve community well-being, but if social capital can be intelligently fostered—
sometimes just by linking several people together into an artificial
social cluster—improvement can be striking.

Take so-called Grameen banking, for example, a brilliant and quite
recent idea from the Bangladeshi economist Muhammad Yunas that
was aimed at helping poor people get loans.[7] Ordinarily, the poor represent a major risk to any lending institution. They find it difficult to
get loans to start a business, go to college, or take any number of steps
that might help to alleviate their poverty. Yunas's idea was to link people together into clusters—if you want to take out a loan, you join a
five-person borrowers' group. The group as a whole then takes responsibility for repaying the loans of any one of its members, and no member can get another loan if the group defaults.

The relationships that develop between the members, some of whom
may have known one another before ever hoping for a loan, are much
stronger and based largely on trust, experience, and shared values.
These links are far stronger and more durable than any individual
would likely have with the lending institution. What's more, the free
and intimate exchange of information among members means that any
good idea one member has for avoiding default and managing to repay
a loan might be used by other members. The tight-knit structure of the
clustered group makes it more than the sum of its parts, and in
Bangladesh, has led to a 97 percent repayment rate on loans.

The importance of social capital has been pointed out many times
before and supported mostly by verbal arguments. As a result, many
economists such as Yunas have begun to take seriously the fact that economic activity is driven not only by individual greedy interest. Crucial
features of social life provide the raw organizational machinery for efficient economic activity. It is at the very least curious that the property
of clustering that arises so naturally in many real-world networks

appears to provide the conditions for the generation of social capital. With the techniques of small-world theory, it may be possible for sociologists or business managers to begin measuring it and perhaps taking measures to promote it.

On the other hand, there are drawbacks to *too much* clustering, as Granovetter's study of job seeking pointed out. To live within a cluster is to be protected from differing norms, and also from truly novel ways of thinking, patterns of behavior, or pieces of information. Sending out an "I need a job" message into a network of friends is not effective, for soon the friends begin hearing the message a second and third time. To reach a large number of people, and especially people with access to information in diverse industries, companies, and regions, one needs to exploit the long-distance links that tie together different clusters. Connectors, people with many weak ties, play the same role, often helping to link together distinct highly connected clusters. As we will see, the absence of such weak ties can have damaging consequences.

NOT SO SMALL CONSEQUENCES

IT HAS LONG been puzzling to sociologists how some communities or organizations respond to crises by mobilizing resources effectively while others do not. Undoubtedly each situation is extremely complicated, with individual personalities and specific and unique circumstances playing a role. But the small-world social architecture—or the lack thereof—may also be an influential factor.

In the early 1960s, for example, Boston's West End was slated for massive demolition in the interests of "urban renewal." In studying the reaction of the largely Italian and working-class community, sociologist Herbert Gans noted, paradoxically, that while the community seemed socially "cohesive" and was in large measure uniformly horrified by the prospects of the demolition, it nevertheless was unable to mobilize in a coordinated way behind local leaders.[8] Gans contrasted this with several other seemingly similar working-class communities that faced similar challenges and managed to organize and act successfully. One of these was also in Boston, in another area known as Charlestown. What might account for the different response in these two areas of Boston?

Gans's original explanation was that working-class culture made members of the community suspicious of "self-seeking" leaders, and so they were loath to join the political organization that might have helped. In his original paper on the strength of weak ties, however, Granovetter pointed to another, perhaps simpler explanation based on a network perspective and the crucial role that weak ties play in binding a community together. Granovetter suggested that in Boston's West End, it could well be that "the neighborhood consisted of cohesive network clusters which were, however, highly decoupled from one another, and that this fragmentation made mobilization difficult, whatever the intentions of individuals."

Indeed, returning to Gans's original description of the event, Granovetter found that local clusters of individuals in fact had managed to mobilize, but the cooperation did not extend throughout the community. It appeared that the diverse subcommunities lacked the connectors, or individuals with weak ties, who would have linked the distinct subcommunities and tied the overall group together. This lack of ties might explain the lack of trust developed in leaders, particularly in leaders rising up from other subcommunities. If you know that a friend of a friend has met and spoken with this leader, there is a short chain connecting the two of you. Such chains tend to make the intentions and motivations of such leaders less suspect because "one has the possibility of exerting influence through the chain in ways that restrain self-seeking."[9]

As Granovetter admitted, it is hard to imagine that there were no weak ties within the community. But Zanette's model for the breakout of a disease may well apply with equal force to the spread of a rumor or information relevant to mobilizing community action. If the density of weak ties were below a threshold, the information might well never break out past a small percentage of the population. If this explanation is right, then the lack of weak ties that would have cemented the community together apparently led to a lack of trust and social capital that potentially could have saved the community. Even if this is not precisely what took place in Boston, the example serves well to make the point: something like this could very easily happen. Small-world effects have real consequences.

Indeed, a very similar scenario may lie behind the success of Silicon Valley as compared with the Route 128 high-technology area in metropolitan Boston. In the 1970s, these two areas were competing more or less equally to be the technological center in the United States. Subsequently, companies in Silicon Valley such as Sun Microsystems,

Hewlett-Packard, and Silicon Graphics emerged as the winners, while their Boston competitors Digital Equipment, Prime Computer, and Apollo Computer were either acquired or dissolved. Why did the high-tech race go one way rather than the other? And did the outcome sever along regional lines? Sociologist Analee Saxenian has argued that some of the most crucial factors were the ease with which ideas, capital, and people flowed not just within single companies but between companies.[10]

Ordinarily, the members of competing organizations share little trust. But in Silicon Valley, different firms seemed willing to cooperate to an extraordinary degree. With employees shifting jobs frequently, those in separate firms often worked together previously, and the culture of computer engineers made technical cooperation and achievement more important than firm loyalty or high salary. This may have been true in Boston as well, but in Silicon Valley, companies readily exploited important interpersonal links that cut across firm boundaries. A freewheeling and open Californian culture, in contrast to a more closed and proprietary nature in New England, seemed to make a huge difference. In Boston, the consequent lack of exchange of ideas and personnel damaged productivity and the ability to move quickly, a deadly shortcoming in the fast-moving high-technology field.

As another example, in the 1970s and 1980s, the Italian automakers Fiat and Alfa Romeo faced similar restructuring in an attempt to reduce costs and improve efficiency. Fiat made the attempt by attacking the influence of the labor unions, stirring up significant conflict with employees, whereas Alfa Romeo managed a more balanced negotiation that helped both labor and management. Alfa Romeo's approach is to be desired, but how did the company manage it? According to sociologist Richard Locke, the outcome can be traced to a difference in the social power structures in Turin and Milan, where the two companies were based.[11]

Locke's studies suggest that in Turin, Fiat's regional base, the political network was highly polarized; it was essentially split into two distinct networks, one associated with business and the other with labor. Each network was well connected within itself, but there were few links connecting the two together. In contrast, the network of political ties in Milan, Alfa Romeo's home, was much more diversely linked, with significant weak ties running between business, labor, and other organizations and associations that could act as intermediaries. This network architecture fostered a happier restructuring for Alfa Romeo in comparison to Fiat.

So the small-world network, in the social case, seems to be a beneficial mix of both clustering and weak links that tie distinct clusters together. Clustering makes for a dense social fabric and allows the formation of social capital, which in turn helps to promote efficiency in decision making. At the same time, weak ties keep everyone close in a social sense to the rest of the community, even if it is very large, which enables each person access to the diverse information and assets of the larger organization. Perhaps organizations and communities should purposefully be built along small-world lines.

Indeed, it begins to seem as if the small-world idea is struggling to express some still deeper insight about how to live in a complex world. At its core lies the idea that too much order and familiarity is just as bad as too much disorder and novelty. We instead need to strike some delicate balance between the two.

SIMPLE WISDOM

IRONICALLY, MUCH OF the work we have explored in this book has been carried out by physicists working in areas that would not normally be considered part of physics: social and computer networks, cellular biochemistry, economics. But in this sense, physics is evolving. Where once it was only the study of matter and the laws of fundamental physics, it has graduated into the study of organization in all its forms. The study of emergence in all its forms is one of the most important scientific enterprises of our era, and will remain that way for the next century. As two eminent physicists recently commented, "The central task of theoretical physics in our time is no longer to write down the ultimate equations but rather to catalogue and understand emergent behavior in its many guises, including potentially, life itself. We call this physics of the next century the study of complex adaptive matter. . . . We are now witnessing a transition from the science of the past, so intimately linked to reductionism, to the study of complex adaptive matter . . . with its hope for providing a jumping-off point for new discoveries, new concepts, and new wisdom."[12]

At the core of this new way of doing science is the perception that the world is in many ways simpler than it appears. Behind the distribution of wealth that stirs up such heated political debate lies not a mess of thousands of competing factors but a simple process of random growth. Mathematically, as it turns out, this process is nearly identical to the way the Internet grows, with the most highly linked sites gathering in new links faster than the rest. Once again, it is also nearly identical to the process by which business firms and cities grow, which is why researchers have found that the distribution of both firms and cities according to size also conforms to Pareto's simple law. Viewed from the proper perspective, many aspects of the world are indeed simpler than they appear.

The small-world idea itself is also remarkably simple. All it takes is a few long-distance links or superconnected hubs, and there you have it—a small world. No doubt this simplicity explains why this kind of network appears in the architecture of everything from the human brain to the web of relationships that bind us into societies, as well as the languages we use to speak and think. Where small-world ideas will lead us in five or ten years is anyone's guess, but they may well reveal something about the way our ideas link up with one another, how discoveries in biology, computer science, sociology, and physics can be so intimately connected, and how studies of Malaysian fireflies can lead, in only a few steps, to new insights across all of science. This too is presumably more than mere coincidence.

NOTES

PRELUDE

1 Henri Poincaré, *La science et l'hypothèse* (Flammarion, Paris, 1902). Introduction.

2 Karl Popper, *The Poverty of Historicism*, p. v (ARK Publishing, London, 1957).

3 Herbert Simon, *Models of My Life*, p. 275 (Basic Books, New York, 1991).

4 John Guare, *Six Degrees of Separation: A Play* (Vintage, New York, 1990).

5 Scientists used to think that each gene carries the instructions for making exactly one protein, but biologists in recent years have found otherwise. The cellular machinery that reads genes and assembles proteins can perform a trick known as *alternative splicing*, in which it selectively ignores or alters the information in a gene when reading it. As a result, one gene can potentially lead to many different proteins. According to Peer Bork of the Max-Delbrück Center for Molecular Medicine in Germany (personal communication), this trick takes place with at least half of all genes, with some leading to the production of over a hundred different proteins (although most produce only two or three).

6 Peter Yodzis, "Diffuse Effects in Food Webs," Ecology, 81, 261–266 (2000).

7 George Joffee, quoted in Mohamad Bazzi, "'A Network of Networks' of Terror," September 16, 2001. Available online at www.newsday.com.

CHAPTER 1. STRANGE CONNECTIONS

1 Ivor Grattan-Guinness, *History of Mathematics* (HarperCollins, London, 1977).

2 Duncan J. Watts and Steven H. Strogatz, "Collective Dynamics of "Small-World" Networks," *Nature* 393, 440–442 (1998).

3 Thomas Blass, "Stanley Milgram: A Life of Inventiveness and Controversy," in G. A. Kimble, C. A. Boneau, and M. Wertheimer, eds., *Portraits of Pioneers in Psychology*, vol. 11 (American Psychological Association, Washington, D.C., 1996).

4 Stanley Milgram, "The Small-World Problem," *Psychology Today* 1, 60–67 (1967).

[5] Stanley Milgram, *Obedience to Authority*, p. 22 (Tavistock Publications, London, 1974).

[6] Ibid., p. 23.

[7] The Internet Movie Database can be found online at www.imdb.com/.

[8] David Kirby and Paul Sahre, "Six Degrees of Monica," *New York Times*, February 21, 1998.

CHAPTER 2. THE STRENGTH OF WEAK TIES

[1] Stanley Milgram, *Obedience to Authority*, p. 30 (Tavistock Publications, London, 1974).

[2] Paul Hoffman, *The Man Who Loved Only Numbers*, p. 7 (Fourth Estate, London, 1998). This is a humorous and delightful account of the life and mathematics of Paul Erdös.

[3] Ibid., p. 6.

[4] Mathematician Jerrold Grossman maintains a Web site devoted to the collaboration graph associated with the small world of Paul Erdös. The address is www.oakland .edu/~grossman/.

[5] Hoffman, *The Man Who Loved Only Numbers*, p. 45.

[6] In mathematical terms, the general result is as follows. In a network with N vertices, the fraction of links required to tie the entire network together into one "giant component" is given by the formula $\ln(N)/N$, where $\ln(N)$ is the natural logarithm of N. This fraction decreases as N gets larger.

[7] Mark Granovetter, "The Strength of Weak Ties," *American Journal of Sociology* 78, 1360–1380 (1973).

[8] Ibid., p. 1373.

[9] Anatol Rapoport and W. Horvath, "A Study of a Large Sociogram," *Behavioral Science* 6, 279–291 (1961).

[10] Mark Granovetter, "The Strength of Weak Ties: A Network Theory Revisited," *Sociological Theory* 1, 203–233 (1983).

[11] George Polya, *How To Solve It* (Princeton University Press, Princeton, 1957).

CHAPTER 3. SMALL WORLDS

[1] John Tierney, "Paul Erdös Is in Town. His Brain Is Open," *Science* 84, p. 40–47 (1984).

[2] H. M. Smith, "Synchronous Flashing of Fireflies," *Science* 82, 151 (1935).

[3] Renato E. Mirollo and Steven Strogatz, "Synchronization of Pulse-Coupled Biological Oscillators," *SIAM Journal of Applied Mathematics* 50, 1645–1662 (1990).

4 Watts and Strogatz were a bit more specific about the number for the "degree of clustering" and for the "number of degrees of separation." Imagine some point in a network—call it X—and think of all the other points to which X is directly linked. In principle, all of these other points could be linked to one another. If every last pair were linked, then the region around X would be as highly "clustered" as it could be. This is akin to all of your friends also being friends with one another, without exception. In reality, in most networks, only a fraction of X's direct neighbors will be linked together. This fraction, a number between 0 and 1, offers a convenient measure of how clustered the region in the vicinity of X is. To measure the degree of clustering for the entire network, you only need to repeat the calculation for every point in turn and then take the average.

For the number of degrees of separation, Watts and Strogatz used a similar definition. Choose two points anywhere in the network, and find how many steps it takes to go between them on the shortest path. This is the "distance" between these points. Again, you can repeat the calculation for every possible pair of points and in the end take the average. This is the number of degrees of separation for the network, the typical number of steps needed to connect two points.

5 One principal issue that remains unanswered is the variation in the natural flashing frequencies within a population of fireflies. It seems that a collection of flies can synchronize more easily if they all tend to flash at the very same frequency. Too much variation in their natural flashing rates can swamp out the forces that would lead to synchronization. Intriguingly, different species appear to have different amounts of variation, which could account for the fact that some synchronize while others do not. See Ivars Peterson, "Step in Time," *Science News* 140, 136–137 (1991).

With regard to hand clapping, the story is even more interesting. Synchronized clapping usually sets in a few seconds after the first wave of clapping greets the end of a performance. Studies reveal that when the synchronization does set in, the frequency of clapping also spontaneously falls by a factor of two—that is, people clap twice as slowly as they did before. Again, the reason seems to be related to variation. Experiments show that slower clappers have less variation in frequency than faster clappers. Indeed, recordings of applause also reveal that after a time of synchronized clapping, people tend to gradually speed up their clapping to make more overall noise. This leads to more variation in the frequencies of all the various clappers, and the synchronization eventually goes away. Crowds have a natural tendency to cycle between episodes of rhythmic and arrhythmic clapping. See Zoltán Néda, Erzsébet Ravasz, Yves Brechet, Tamás Vicsek, and Albert-Lászlo Barabási, "Self-Organizing Processes: The Sound of Many Hands Clapping, *Nature* 403, 849–850 (2000).

6 Duncan J. Watts and Steven H. Strogatz, "Collective Dynamics of 'Small-World' Networks," *Nature* 393, 440–442 (1998).

CHAPTER 4. BRAIN WORKS

1 Manfred Eigen, *The Physicist's Conception of Nature*, ed. Jagdish Mehra (Reidel, Dordrecht, 1973).

2 From a letter written in 1798 by Franz Joseph Gall to his friend Joseph von Retzer. See the excellent Web site, The History of Phrenology on the Web, maintained by John van Wyhe (www.jmvanwyhe.freeserve.co.uk).

3 Ibid.

4 J. W. Scannell, "Determining Cortical Landscapes," *Nature* 386, 452 (1997).

5 Vito Latora and Massimo Marchiori, "Efficient Behavior of Small-World Networks," *arXiv: cond-mat*/0101396, January 25, 2001. (This paper and a number of others cited in this book are available online at the physics community's "pre-print" archive, currently maintained by Los Alamos National Laboratory and located at http://xxx .lanl.gov/. This paper, for example, can be retrieved from the address http://xxx.lanl .gov/cond-mat/0101396. Henceforth, I shall refer to papers of this sort with the abbreviated reference "*arXiv*," which refers to this resource.)

6 Miguel Castelo-Branco, Rainer Goebel, Sergio Neuenschwander, and Wolf Singer, "Neural Synchrony Correlates with Surface Segregation Rules," *Nature* 405, 685–689 (2000). See also Marina Chicurel, "Windows on the Brain," *Nature* 412, 266–268 (2001).

7 Kate MacLeod, Alex Bäcker, and Gilles Laurent, "Who Reads Temporal Information Contained across Synchronized and Oscillation Spike Trains?," *Nature* 395, 693 (1998).

8 Luis F. Lago-Fernández, Ramón Huerta, Fernando Corbacho, and Juan A. Sigüenza, "Fast Response and Temporal Coding on Coherent Oscillations in Small-World Networks," *arXiv: cond-mat*/9909379, September 27, 1999.

9 Socrates, quote is from Anthony Gottlieb, *The Dream of Reason*, p. 28 (Allen Lane, London, 2000).

10 Eric Temple Bell, quote from Alan Mackay, *A Dictionary of Scientific Quotations*, p. 25 (IOP Publishing, London, 1991).

CHAPTER 5. THE SMALL-WORLD WEB

1 Gerard Bricogne, quote from Alan Mackay, *A Dictionary of Scientific Quotations*, p. 39 (IOP Publishing, London, 1991).

2 William J. Jorden, "Soviets Claiming Lead in Science," *New York Times*, October 5, 1957.

3 Katie Hafner and Matthew Lyon, *Where Wizards Stay Up Late: The Origins of the Internet*, p. 22 (Touchstone, New York, 1996).

4 Paul Baran, "Introduction to Distributed Communications Networks," report RM-3420-PR (RAND Corporation, Santa Monica, Calif., August 1964). This report is available online at www.rand.org/publications/RM/baran.list.html.

5 Hafner and Lyon, *Where Wizards Stay Up Late*.

6 Bernardo Huberman, Peter Pirolli, James Pitkow, and Rajan Lukose, "Strong Regularities in World Wide Web Surfing, *Science* 280, 95 (1998). The Web site for the Xerox Internet Ecologies Division of the Palo Alto Research Center is also worth a visit: www.parc.xerox.com/istl/groups/iea/dynamics.shtml.

7 Peter Drucker, "Beyond the Information Revolution," *Atlantic Monthly* 284, 47–57 (October 1999).

8 "IBM's Gerstner Speaks on e-Commerce," *Newsbytes News Network,* March 19, 1998.

9 Les Alberthal, "The Once and Future Craftsman Culture," in Derek Leebaert, ed., *The Future of the Electronic Marketplace* (MIT Press, Cambridge, 1998).

10 Peter Fingar, Harsha Kumar, and Tarun Sharma, "21st Century Markets: From Places to Spaces," *First Monday* 4 (December 1999). Available online at http://firstmonday.org /issues/issue4_12/fingar/index.html.

11 Michael Faloutsos, Petros Faloutsos, and Christos Faloutsos, "On Power-Law Relationships of the Internet Topology," *Computer Communication Review* 29, 251 (1999).

12 See Réka Albert and Albert-László Barabási, "Statistical Mechanics of Complex Networks," *Reviews of Modern Physics,* in press.

13 As a technical point, the Faloutsos study looked predominantly at the Internet at what is called the "interdomain" level. Whether you are an independent user or part of some organization, your computer links originally to a computer known as a "router." This router and all the computers it serves make up a local area network. This router itself is then linked into a network of other routers in what is known as a "domain." A domain, in effect, is a network of routers. However, at least some of the routers within a domain will also be linked to routers in other domains. As a consequence of this, you can study the Internet structure at two different levels: you might study the way routers are linked together within a domain (the router level) or the way domains are linked together (the interdomain level). The distinction is not terribly important for our purposes, however. In fact, the Faloutsos brothers found similar patterns in either case.

14 Réka Albert, Hawoong Jeong, and Albert-László Barabási, "Diameter of the World Wide Web," *Nature* 401, 130–131 (1999).

15 Hawoong Jeong, Balint Tombor, Réka Albert, Zoltan N. Oltvai, and Albert-László Barabási. "The Large-Scale Organization of Metabolic Networks," *Nature* 407, 651–654 (2000).

16 Sidney Redner, "How Popular Is Your Paper?" *European Physics Journal B* 4, 131 (1998); Mark E. J. Newman, "The Structure of Scientific Collaboration Networks," *Proceedings of the National Academy of Sciences of the United States of America* 98, 404–409 (2001).

17 Ramon Ferrer i Cancho and Ricard V. Solé, "The Small World of Human Language," working paper 01-03-016, Sante Fe Institute, Sante Fe, N.M., 2001.

CHAPTER 6. AN ACCIDENTAL SCIENCE

[1] Leucippus, quote from Alan Mackay, *A Dictionary of Scientific Quotations*, p. 152 (IOP Publishing, London, 1991).

[2] Fustel de Coulanges, quote from Fritz Stern, ed., *The Varieties of History: From Voltaire to the Present*, p. 57 (World Publishing, Cleveland, 1956).

[3] Ralph H. Gabriel, *American Historical Review* 36, 786 (1931).

[4] Things are not quite so simple, however, as the proponents of a theory can always discount the evidence and go on arguing, or modify the theory in some way rather than reject it. But experiments do offer the opportunity to place considerable additional stress on theories, and to subject them to well-conceived tests. For an excellent account of the real-world difficulties in judging theories on experimental evidence, see Harry Collins and Trevor Pinch, *The Golem* (Cambridge University Press, Cambridge, 1993).

[5] Stephen Jay Gould, *Wonderful Life*, p. 283 (Hutchinson Radius, London, 1989).

[6] Lord Rayleigh's reasoning fails in the following way: In any liquid, molecules form weak bonds with their neighbors. A molecule buried inside the liquid finds itself in a good situation—it can form bonds with molecules on all sides. But molecules at the surface are not so lucky. Unable to form bonds on one side, these surface dwellers end up being unsatisfied, and as a result have higher energy than their buried cousins. The extra cost in energy of an exposed surface gives it a natural tension—it tends to shrink, in so far as possible. This "surface tension" is why water forms spherical droplets, which have as little surface area as possible. In Bénard's experimental setup, the liquid surface has little opportunity to shrink, but it can suck up warm liquid from the depths and use it to replace the cold liquid at the surface. As it turns out, the surface tension of a liquid decreases with increasing temperature, so bringing up warm liquid reduces the surface tension. As a result, it reduces the extra energy of the surface. In Bénard's experiment, as scientists now understand, surface tension, not bouyant warm water rising, as Rayleigh had supposed, actually drives the fluid flow.

[7] Leon Trotsky, quote from Edward Hallett Carr, *What Is History?*, p. 102 (Penguin, London, 1990).

[8] Ignacio Rodríguez-Iturbe and Andrea Rinaldo, *Fractal River Basins* (Cambridge University Press, Cambridge, 1997).

CHAPTER 7. THE RICH GET RICHER

[1] Mark Twain. See Alan Mackay, *A Dictionary of Scientific Quotations*, p. 244 (IOP Publishing, London, 1991).

[2] "Eight Injured in Bradford Riots," *The Guardian (London)*, April 16, 2001.

[3] Mark Granovetter, "Threshold Models of Collective Behavior," *American Journal of Sociology* 83, 1420–1443 (1978). This simple example of thresholds may seem somewhat artificial. Granovetter also investigated the case in which the thresholds for people

in a group are clustered about some average value, say, 25, and looked at how the width or spread in these values would affect group behavior. Under these much more realistic assumptions, he still discovered surprising and unexpected behavior. When the spread was fairly narrow, no more than about six people would join the riot. As the spread grew larger, however, there was a dramatic shift in the group dynamic, and most of the one hundred people would now become involved. This more realistic framework illustrates how changes in apparently insignificant features can have startling repercussions.

4 Steve Lawrence and C. Lee Giles, "Accessibility of Information on the Web," *Nature* 400, 107 (1999).

5 Albert-László Barabási and Réka Albert, "Emergence of Scaling in Random Networks," *Science* 286, 509–512 (2001).

6 Fredrik Liljeros, Christofer Edling, Luís Nunes Amaral, H. Eugene Stanley, and Yvonne Åberg, "The Web of Human Sexual Contacts," *Nature* 411, 907–908.

7 Malcolm Gladwell, *The Tipping Point*, p. 36 (Little Brown, New York, 2000).

8 Liljeros et al., "The Web of Human Sexual Contacts."

9 This power-law distribution is special in that there is no "typical" number of links. In other words, the network has no inherent bias to produce elements with an expected number of links; rather this number varies widely over a huge range. That is to say, there is no inherent "scale" for the number of links, and the network is scale-free.

10 Solomon Asch, "Effects of Group Pressure upon the Modification and Distortion of Judgement," in H. Guetzkow, ed., *Groups, Leadership, and Men* (Carnegie Press, Pittsburgh, 1951).

11 Irving Janus, *Groupthink* (Houghton Mifflin, Boston, 1982).

12 Mark Newman, "Clustering and Preferential Attachment in Networks," *Physical Review E* 64, 25102 (2001).

13 Hawoong Jeong, Zoltan Neda, and Albert-László Barabási, "Measuring Preferential Attachment for Evolving Networks," *arXiv: cond-mat*/0104131, April 14, 2001.

14 Gerald F. Davis, Mina Yoo, and Wayne E. Baker, "The Small World of the Corporate Elite," preprint, University of Michigan Business School, Ann Arbor, 2001.

15 Mark S. Mizruchi, "What Do Interlocks Do? An Analysis, Critique and Assessment of Research on Interlocking Directorates," *Annual Review of Sociology* 22, 271–298 (1996).

CHAPTER 8. COSTS AND CONSEQUENCES

1 D'Arcy Wentworth Thomson, *On Growth and Form* (Cambridge University Press, London, 1917).

2 Nebojša Nakićenović, "Overland Transportation Networks: History of Development and Future Prospects," in David Batten, John Casti, and Roland Thorn, eds., *Networks in Action* (Springer-Verlag, Berlin, 1995).

3 Cynthia Barnhart and Stephane Bratu, "National Trends in Airline Flight Delays and Cancellations" (presentation at the Workskop on Airline and National Strategies for Dealing with Airport and Airspace Congestion, University of Maryland, March 15–16, 2001). Available online at www.isr.umd.edu/airworkshop/Barnhart-Bratu.pdf.

4 Jon Hilkevitch, "FAA Says Some of the Flak It Takes Is Right on Target," *Chicago Tribune*, July 18, 2001.

5 George L. Donohue, Testimony before the U.S. House of Representatives Committee on Appropriations Subcommittee on Transportation, March 15, 2001. Available online at www.isr.umd.edu/airworkshop/Donohue2.pdf.

6 Keith Harper, "Britain's Crowded Skies Most Dangerous in Europe," *Guardian*, July 23, 2001.

7 M. Hanson, H. S. J. Tsao, S. C. A. Huang, and W. Wei, "Empirical Analysis of Airport Capacity Enhancement Impact: Case Study of DFW Airport" (presented at the 78th annual meeting of the National Research Council Transportation Research Board, January 10–14, 1999).

8 Luís A. Nunes Amaral, Antonio Scala, Marc Barthélémy, and H. Eugene Stanley, "Classes of Behaviour of Small-World Networks," *arXiv: cond-mat*/0001458, January 31, 2000.

9 Peter Fitzpatrick, "Aircraft Industry Flying High after Bombadier Deal," *Financial Post* (Canada), July 10, 2001.

10 Soon-Hyung Yook, Hawoong Jeong, and Albert-László Barabási, "Modeling the Internet's Large-Scale Topology," *arXiv:cond-mat*/0107417, July 19, 2001.

11 Office of the President of the United States, *A National Security Strategy of Engagement and Enlargement* (White House, Washington, D.C.,1996). Available online at www.fas.org/spp/military/docops/national/1996stra.htm.

12 *Presidential Decision Directive 63. The Clinton Administration's Policy on Critical Infrastructure Protection* (White House, Washington, D.C., May 22, 1998). Available online at www.cybercrime.gov/white_pr.htm.

13 Bill Gertz, "Computer Hackers Could Disable Military," *Washington Times*, April 16, 1998.

14 Steve Gibson has written a chilling account of the episode, which is available on his Web site, www.GRC.com.

15 John M. Deutch, Office of the Director of Central Intelligence, Foreign Information Warfare Programs and Capabilities (speech before the Senate Subcommittee on Intelligence, June 25, 1996). Available online at www.odci.gov/cia/public_affairs/speeches /archives/1996/dci_testimony_062596.html.

16 Robert H. Anderson et al., "Securing the Defense Information Infrastructure: A Proposed Approach," p. ii, report of the RAND National Defense Research Institute, RAND Corp., Santa Monica, 1999. Available online at http://graylit.osti.gov/.

[17] Colonel Daniel J. Busby, "Peacetime Use of Computer Network Attack, U.S. Army War College Strategy Research Project" (U.S. Army War College, Carlyle, 1999). Available online at http://graylit.osti.gov/.

[18] It seems likely that a network of the egalitarian type would be similarly robust but for a slightly different reason. In such a network, it is the few long-distance links that make the network small, and very few elements are involved in such links. Hence, a random attack would fail to knock out these bridges and would leave the network largely intact.

[19] Réka Albert, Hawoong Jeong, and Albert-László Barabási, "Error and Attack Tolerance of Complex Networks," Nature 406, 378–381 (2000).

[20] Anderson et al., "Securing the Defense Information Infrastructure."

[21] Institute of Medicine, Emerging Infections: Microbial Threats to Health in the United States (National Academy Press, Washington, D.C., 1992).

[22] Mitchell L. Cohen, "Changing Patterns of Infectious Disease," Nature 406, 762–767 (2000).

[23] J. S. Edwards and Bernhard Palsson, "The Escherichia coli MG1655 in silico Metabolic Genotype: Its Definition, Characteristics and Capabilities," Proceedings of the National Academy of Sciences of the United States of America 97, 5528–5533 (2000).

[24] Hawoong Jeong, Sean Mason, Albert-László Barabási, and Zoltan Oltvai, "Lethality and Centrality in Protein Networks," Nature 411, 41–42 (2001).

CHAPTER 9. THE TANGLED WEB

[1] Max Gluckman, Politics, Law and Ritual (Mentor Books, New York, 1965).

[2] Doug Struck, "Japan Blames Whales for Lower Fish Catch," International Herald Tribune, July 28–29, 2001.

[3] Jeremy B. C. Jackson et al., "Historical Overfishing and the Recent Collapse of Coastal Ecosystems," Science 293, 629–638 (2001).

[4] Ibid., p. 635.

[5] United Nations Food and Agriculture Organization, "Review of the State of World Marine Fishery Resources," technical paper 335, UN Food and Agriculture Organization, New York, 1994.

[6] Leslie Harris, "Independent Review of the State of the Northern Cod Stock," final report, Northern Cod Review Panel, prepared for the Honourable Thomas Siddon, Minister of Fisheries, 1990.

[7] For an excellent review of the controversy, see David M. Lavigne, "Seals and Fisheries, Science and Politics" (talk given at the Eleventh Biennial Conference on the Biology of Marine Mammals, Orlando, Fla., December 14–18, 1995). Available online at www.imma.org/orlando.pdf.

[8] Jeffrey A. Hutchings and Ransom A. Myers, "What Can Be Learned from the Collapse of a Renewable Resource? Atlantic Cod, *Gadus morhua,* of Newfoundland and Labrador," *Canadian Journal of Fisheries and Aquatic Sciences* 51, 2126–2146 (1994).

[9] S. Strauss, "Decimated Stocks Will Recover if Fishing Stopped, Study Finds. East Coast Decline in Cod Resulted from Overfishing, Not Seals," *Globe and Mail,* August 25, 1995.

[10] In 1995, a group of ninety-seven marine biologists from all over the world signed a petition objecting to the Canadian government's actions: "As professionals in the field of marine mammal biology we disagree with the Canadian government's statement that North Atlantic seals are a 'conservation problem'. All scientific efforts to find an effect of seal predation on Canadian groundfish stocks have failed to show any impact. Overfishing remains the only scientifically demonstrated conservation problem related to fish stock collapse." *Comment on Canada's Seal Policy,* signed by biologists at the Eleventh Biennial Conference on the Biology of Marine Mammals, Orlando, Fla., December 14–18, 1995. See www.imma.org/petition.html.

[11] S. D. Wallace and J. W. Lawson. "A Review of Stomach Contents of Harp Seals *(Phoca groenlandica)* from the Northwest Atlantic: An Update," technical report 97-01, International Marine Mammal Association, Guelph, Ontario, Canada, 1997. See also "Report of the International Scientific Workshop on Harp Seal-Fishery Interactions in the Northwest Atlantic, 24–27 February 1997, St. John's, Newfoundland," p. 37, International Marine Mammal Association, Guelph, Ontario, Canada, 1997. Available online at www.imma.org/workshop.pdf.

[12] Ten million is a rough estimate. Peter Yodzis has made a more accurate estimate for the food web of the Benguela ecosystem off the coast of South Africa. Here fisheries have demanded culls of the Cape fur seal, claiming that they eat too many hake. But Yodzis found that the number of links between Cape fur seals and hake involving no more than eight species was 28,722,675. Peter Yodzis, "Diffuse Effects in Food Webs," *Ecology* 81, 261–266 (2000).

[13] Peter Yodzis, "The Indeterminacy of Ecological Interactions, as Perceived through Perturbation Experiments," *Ecology* 69, 508–515 (1988).

[14] Charles Elton, quoted from Kevin McCann, "The Diversity-Stability Debate," *Nature* 405, 228–233 (2000).

[15] Robert M. May, *Stability and Complexity in Model Ecosystems* (Princeton University Press, Princeton, 1973).

[16] Peter Yodzis, "The Stability of Real Ecosystems," *Nature* 289, 674–676 (1981).

[17] David Tilman and John A. Downing, "Biodiversity and Stability in Grasslands," *Nature* 367, 363–365 (1994).

[18] Stuart L. Pimm, John H. Lawton, and Joel E. Cohen, "Food Web Patterns and Their Consequences," *Nature* 350, 669–674 (1991).

[19] Kevin McCann, Alan Hastings, and Gary Huxel, "Weak Trophic Interactions and the Balance of Nature," *Nature* 395, 794–798 (1998).

[20] The 97-hectare plot associated with Scotch broom is just one of many experimental ecosystems under study at Silwood Park.

[21] Ricard Solé and José Montoya, "Complexity and Fragility in Ecological Networks," working paper 00-11-060, Santa Fe Institute, Sante Fe, N.M., 2000. Available online at www.santafe.edu/sfi/publications/00wplist.html.

[22] Richard J. Williams, Neo D. Martinez, Eric L. Berlow, Jennifer A. Dunne, and Albert-László Barabási, "Two Degrees of Separation in Complex Food Webs," working paper 01-07-036, Santa Fe Institute, Santa Fe, N.M., 2001. Available online at www.santafe.edu/sfi/publications/01wplist.html.

[23] Stuart Pimm and Peter Raven, "Extinction by Numbers," *Nature* 403, 843–844 (2001).

[24] McCann, "The Diversity-Stability Debate."

CHAPTER 10. TIPPING POINTS

[1] Fyodor Dostoyevsky, *Notes from Underground*, p. 18 (Penguin, London, 1972).

[2] The story of Yuri Rumer, Moissey Koretz, and Lev Landau and their troubles with the NKVD can be found in Gennady Gorelik, "The Top Secret Life of Lev Landau," *Scientific American*, 277, no. 2, 72–77 (August 1997).

[3] Landau's explanation later won him the Nobel Prize. He showed how the laws of quantum theory turn liquid helium at low temperatures into a "superfluid," a bizarre new liquid form of matter that lacks any trace of internal friction. A superfluid set swirling in a cup will swirl forever, never coming to rest.

[4] Malcolm Gladwell, *The Tipping Point*, p. 7 (Little, Brown, New York, 2000).

[5] Financier Bernard Baruch, quote from Robert Prechter Jr., *The Wave Principle of Human Social Behaviour* (New Classics Library, Gainesville, 1999).

[6] Robert J. Shiller, *Irrational Exuberance*, pp. 177–178. (Princeton University Press, Princeton, 2000),

[7] Richard Dawkins, *The Selfish Gene*, p. 209 (Oxford University Press, Oxford, 1976).

[8] Seth Godin, *Unleashing the Ideavirus* (Do You Zoom, Dobbs Ferry, N.Y., 2000).

[9] Peter Beilenson et al., "Epidemic of Congenital Syphilis—Baltimore, 1996–1997," *Morbidity and Mortality Weekly Report* 47, 904–907 (1998).

[10] Peter Beilenson et al., "Outbreak of Primary and Secondary Syphilis—Baltimore City, Maryland, 1995," *Morbidity and Mortality Weekly Report* 45, 166–169 (1996).

[11] John Potterat, quote from Gladwell, *Tipping Point*, p. 17.

[12] There is a fine point. Physicists actually know that phase transitions come in two general kinds: so-called continuous and discontinuous phase transitions. Landau's theory was meant to describe just one of these classes, still an immense range of phenomena.

[13] The technical details of the universal theory of organizational transformation are not easy, but an excellent source that makes things about as clear as possible is James J. Binney, Nigel Dowrick, Andrew Fisher, and Mark Newman, *The Theory of Critical Phenomena* (Oxford University Press, Oxford, 1992).

[14] Haye Hinrichsen, "Critical Phenomena in Nonequilibrium Systems," *Advances in Physics* 49, 815–958 (2000).

CHAPTER 11. BREAKING OUT, SMALL-WORLD STYLE

[1] Jonathan Mann, quote from the preface to Laurie Garrett, *The Coming Plague*, p. xv (Penguin Books, New York, 1994).

[2] William H. Stewart, quote from Mitchell Cohen, "Changing Patterns of Infectious Disease," *Nature* 406, 762–767 (2000).

[3] World Health Organization, *Removing Obstacles to Healthy Development* (World Health Organization, Geneva, 1999).

[4] UNAIDS, Joint United Nations Programme on HIV/AIDS, AIDS Epidemic Update: December 2000. Available online at www.unaids.org/wac/2000/wad00/files/WAD_epidemic_report.htm.

[5] Mann, quote from the preface to Garrett, *Coming Plague*, p. xv.

[6] Edward Hooper, *The River* (Penguin Books, London, 2000).

[7] For a diagram of some of the early animal movements that spread the disease, see www.maff.gov.uk/.

[8] Damián Zanette, "Critical Behaviour of Propagation on Small-World Networks," *arXiv: cond-mat*/0105596, May 30, 2001.

[9] Simon Wain-Hobson, quote from Josie Glausiusz, "The Year in Science: The Chasm in Care," *Discover* 20, 40–41 (January 1999).

[10] Norman Miller and Roger Yeager, "By Virtue of Their Occupation, Soldiers and Sailors Are at Greater Risk," *AIDS Analysis Asia* 1, 8–9 (1995).

[11] Romualdo Pastor-Satorras and Alessandro Vespignani, "Epidemic Spreading in Scale-Free Networks," *Physical Review Letters* 86, 3200–3203 (2001).

[12] Romualdo Pastor-Satorras and Alessandro Vespignani, "Optimal Immunisation of Complex Networks," *arXiv: cond-mat*/0107066, July 3, 2001.

[13] Pastor-Satorras and Vespignani, "Optimal Immunisation of Complex Networks." See also Zoltán Dezsö and Albert-László Barabási, "Can We Stop the AIDS Epidemic?" *arXiv: cond-mat/0107420*, July 19, 2001.

[14] For a brief review, see William W. Darrow, John Potterat, Richard Rothenberg, Donald Woodhouse, Stephen Muth, and Alden Klovdahl, "Using Knowledge of Social Networks to Prevent Human Immunodeficiency Virus Infections: The Colorado Springs Study," *Sociological Focus* 32, 143–158 (1999).

CHAPTER 12. LAWS FOR THE LIVING

[1] Robert Laughlin and David Pines, "The Theory of Everything," *Proceedings of the National Academy of Sciences of the United States of America* 97, 28–31 (2000).

[2] *The Concise Oxford Dictionary,* 6th ed. (Oxford University Press, Oxford, 1976).

[3] Thomas Schelling, "Dynamic Models of Segregation," *Journal of Mathematical Sociology* 1, 143–186 (1971).

[4] Adam Smith, *An Enquiry into the Nature and Causes of the Wealth of Nations,* book I, chapter 2 (A. Strahan, London, 1776).

[5] Richard Thaler, *The Winner's Curse* (Princeton University Press, Princeton, 1994).

[6] Of course, one can argue about the definition of *works*. Most economists at the moment measure economic health by economic growth as measured by gross domestic product. But this approach leaves out many factors that ought to be measured in social well-being, such as a clean environment, low unemployment, affordable housing, and so on. Economics as currently practiced is decidedly skewed toward matters that can be measured in strictly financial terms.

[7] See, for example, Robert Shiller, *Irrational Exuberance* (Princeton University Press, Princeton, 2000), and Andrei Shleifer, *Inefficient Markets* (Oxford University Press, Oxford, 2000).

[8] For a sample of these fascinating models, see Jean-Philippe Bouchaud and Rama Cont, "Herd Behavior and Aggregate Fluctuations in Financial Markets," *arXiv: cond-mat/9712318,* December 30, 1997; Thomas Lux and Michele Marchesi, "Scaling and Criticality in a Stochastic Multi-Agent Model of a Financial Market," *Nature* 397, 498–500 (1999); Damian Challet, Alessandro Chessa, Matteo Marsili, and Yi-Chen Zhang, "From Minority Games to Real Markets," *Journal of Quantitative Finance* 1, 168 (2001).

[9] Vilfredo Pareto, *Cours d'economique politique* (Macmillan, London, 1897). Strictly speaking, the pattern of wealth distribution discovered by Pareto does not hold across the board, but applies with increasing accuracy toward the wealthy end of the distribution. However, this does not affect any of the conclusions that follow.

[10] Wataru Souma, "Universal Structure of the Personal Income Distribution," *arXiv: cond-mat/0011373,* November 22, 2000.

[11] John Kenneth Galbraith, *A History of Economics,* pp. 6–7 (Penguin, New York, 1987).

[12] Jean-Philippe Bouchaud and Marc Mézard, "Wealth Condensation in a Simple Model of Economy," *Physica A* 282, 536 (2000).

[13] John Flemming and John Micklewright, "Income Distribution, Economic Systems and Transition," *Innocenti Occasional Papers, Economic and Social Policy Series,* no. 70 (UNICEF International Child Development Centre, Florence, 1999).

CHAPTER 13. BEYOND COINCIDENCE

1 Quote from Alan Mackay, *A Dictionary of Scientific Quotations*, p. 133 (IOP Publishing, London, 1991).

2 Herbert Simon, *Models of My Life*, p. 275 (Basic Books, New York, 1991).

3 Mark Granovetter, "Economic Action and Social Structure: The Problem of Embeddedness," *American Journal of Sociology* 91, 481–510 (1985).

4 Mark Granovetter, "A Theoretical Agenda for Economic Sociology," in Mauro F. Guillen, Randall Collins, Paula England, and Marshall Meyer, eds., *Economic Sociology at the Millennium* (Russell Sage Foundation, New York, 2001).

5 Stanley Milgram, *Obedience to Authority*, p. 151 (Tavistock, London, 1974).

6 Francis Fukayama, *Trust*, p. 26 (Free Press Paperbacks, New York, 1995).

7 Muhammad Yunas, "The Grameen Bank," *Scientific American* 281, 114–119 (November 1999).

8 Herbert Gans, *The Urban Villagers* (Free Press, New York, 1962).

9 Granovetter, "A Theoretical Agenda for Economic Sociology."

10 Analee Saxenian, *Regional Advantage: Culture and Competition in Silicon Valley and Route 128* (Harvard University Press, Cambridge, 1994).

11 Richard Locke, *Remaking the Italian Economy* (Cornell University Press, Ithaca, N.Y., 1995).

12 Robert Laughlin and David Pines, "The Theory of Everything," *Proceedings of the National Academy of Sciences of the United States of America* 97, 28–31 (2000).

INDEX

Page numbers in *italics* refer to illustrations.

actors' networks, 28–30, 56
  name recognition and, 113
  Oracle of Kevin Bacon, 28–29
  preferential attachment in, 111,
    113
Adams, Edie, 29
Advanced Research Project Agency
    (ARPA), 74–76
advertising, 160–61
Afghanistan, 21
Africa, AIDS in, 171, 172–75, 178–80
AIDS, 22, 33, 137, 171–75, 178–80,
    184
  earliest known case of, 174
  effective noninfectiousness of,
    176–77
  natural transfer of, 173–74
  origins of, 172–75, 178–79
  polio vaccines and, 174
  prevalence of, 171
  reused syringes and, 179
  sociological changes and, 178–79,
    180
  soldiers as targets of, 179
  treatment strategies for, 181–82,
    183

air traffic control, 122, 128, 132
air transportation networks, 121–25
  airport hubs of, 114, 121–22,
    123–24, 125
  global, 172, 175
  maximum capacity of, 122
  number of runways and, 122–23
  regional airlines of, 125
Alberthal, Les, 77
Albert, Réka, 87, 111–13, 119, 124,
    126, 130–33, 198
Alfa Romeo, 206–7
Algeria, 21
alternative splicing, 209*n*
Amaral, Luis, 124
Amazon.com, 97, 123
  "denial of service" smurf attack
    on, 128
Amin, Idi, 129
antibiotics, 161, 170, 171
  new, 134
Arctic rocks, patterns made by, 95, *96*
aristocratic networks, *112*, 118–20,
    123, 124–25, 126, 137
  failure of, 130, 131–32
  food webs as, 149–52

aristocratic networks *(continued)*
 hubs in, 119, 123, 124, 131–32
 Internet as, 125, 130, 131
 living cell as, 134
 spread of disease in, 180–82, 183
Army War College, U.S., 129
ARPANET, 75–76
Asch, Solomon, 115
Atlantic cod, 139, 140–43
ATP (adenosine triphosphate), 134
audience, synchronized clapping by,
 49, 211*n*
Australia, 32
authoritative relationships, 26–27,
 200–201
automakers, Italian, 206–7
axons, 64, 65

Bacon, Kevin, 28–29, 34–35
bacteria, 137
 drug-resistant, 134–35, 171
 E. coli, 87, 97, 133, 134, 135, 171
 S. aureus, 134
 T. palladium, 162
Balkans, 159
Baltimore, Md., 161–62, 164
Bangladesh, 203
banking, 77, 128, 129
 Grameen, 203
Barabási, Albert-László, 83–86, 87,
 111–13, *112*, 119, 124, 126,
 130–33, 134, 136, 198
Baran, Paul, 74–75, 78–80, *79*, 81, 82,
 86
Baruch, Bernard, 159–60
Becker, Carl, 89
behavioral economics, 188
Belgian Congo, 174
Bell, Alexander Graham, 76
bell curve, statistical, 84, 189
Bell, Eric Temple, 71
Bénard, Henri, 92–93, 94, 96, 97,
 214*n*
Ben Ghaln, Salah, 13–14
Bingenheimer, Rodney, 29

bin Laden, Osama, 21
biomass, 146
Birch, Hal, 80–81, 82
Borneo, 59
Boston, Mass., 204–6
 Route 128 technological center of,
 205–6
Boston University, 113–14, 124
Bouchaurd, Jean Philippe, 191–96
Bradford, England, riot at, 106–7
brain:
 locust, 68, 69–70
 mammalian, 64–65, 66, 67–68
brain, human, 12, 15, 59, 61–72, 137,
 197–98, 208
 Broca's area of, 64, 66
 cerebral cortex of, 63, 64–65
 consciousness and, 66–69, 70
 degree of clustering in, 65–66
 degree of separation in, 65
 as egalitarian network,
 119, 125
 Gall's theory of, 61–63, 212*n*
 hippocampus of, 63, 64, 66
 information storage in, 70
 modules of, 61–64
 neural networks of, 64–66, 67
 neural synchrony of, 49,
 67–70
Brando, Marlon, 13–14
Brenner, Sidney, 59
brewer's yeast, genome of, 135–37,
 136
Bricogne, Gerard, 73
British Airways, 122
British National Corpus, 88
Broca, Paul, 64
Broca's area, 64, 66
Brown, Courtney, 29
business networks, 15, 21, 22, 88
 corporate boards of directors as,
 116–18
 management networks of, 19
 spreading information in, 117
Business Traveler, 123

Caenorhabditis elegans, 14–15, 59–60, 119, 125
Call Me Bwana, 29
Cambridge University, 21
Canada, fishing industry of, 140–41, 143, 153, 218*n*
cardiac pacemaker cells, 49, 50
catchment areas, 99
cats, brain of, 64–65, 67–68
Cedar Creek National History Area, 145–46
cell, living, 12, 15, 19, 87, 91, 96, 97, 105, 133–37
  as aristocratic network, 134
  degree of separation in, 135
  genes of, 133, 209*n*
  see also proteins
Centers for Disease Control, 162, 163–64, 170
Central Intelligence Agency, 129
cerebral cortex, 63, 64–65
Cheers, 30
Cheswick, Bill, 80–81, 82
chimpanzees, 173, 174
Clinton, Bill, 30, 114
  security threats foreseen by, 127–28
clocks, synchronized pendulums of, 49–50
CNN, 128
cod, Atlantic, 139, 140–43
Cold War, 11, 73–76
Coleman, James, 201
"Collective Dynamics of 'Small-World' Networks" (Watts and Strogatz), 23–24
collective unconscious, 40
Colorado Springs, 183
communism, 11
complex adaptive matter, study of, 207–8
complexity theory, 18–22, 127
computation, 59–60
computer graphics, 32
computer modeling:

of cyber-crime, 128–29, 130–32
of disease spreading, 177–80, 205
of distribution of wealth, 192–96
of fireflies' synchronous flashing, 50, 58–59
of locust's neural synchrony, 69–70
of river networks, 101–3, *102*
of small-world graphs, 53–55
of World Wide Web, 86
computer networks, 21, 126–33
  failure of, 130–33
  random, 130–31
  redundancy of, 130, 131, 132
  of U.S. Defense Department, 128–29
  see also cyber-crime; Internet; World Wide Web
computer viruses, 128
concurrence-seeking, 115–16
Congress, U.S., 122, 129
connectors, 120, 204
  as core group, 182–83
  in ecosystems, 149, 151–52, 153–54
  as hubs, 114–15, 119, 149
  sexually transmitted diseases and, 180–83
  tipping point and, 168–69
Connick, Harry, Jr., 28
consciousness, 66–69, 70
conservation, ecological, 152
contact process games, 166–68, *167*
core groups, 182–83
corporate boards of directors, networks of, 116–18
crack cocaine, 161–62, 164
crickets, synchronous chirping of, 49, 50, 68
crime rate, 159
critical phenomena, theory of, 166
Crosby, Bing, 28–29
cross-species infections, 173–74
cyber-crime, 127–33
  of adolescent hackers, 128, 129
  computer simulations of, 128–29, 130–32

cyber-crime *(continued)*
  computer viruses, 128
  coordinated vs. uncoordinated,
      129–30, 131–33, 135
  smurf attacks, 128–29

Danson, Ted, 30
Darrow, William, 183
Darwin, Charles, 91–92, 146
data encryption, 132
Davis, Donald, 75
Davis, Gerald, 117–18
Dawkins, Richard, 160
de Coulanges, Fustel, 89
Defense Department, U.S., 128–29
degree of clustering, 55, 82, 91, 201,
      202–4, 211n
  in actors' networks, 56
  in brain, 65–66
  of English language, 88
  excessive, 204–7
  in friendship networks, 38–40, 65
  in Internet, 82
  of nematode neural network, 59–60
  of ordered networks, 51, 53–54
  in preferential attachments, 113
  of random networks, 38–40,
      51–52, 54, 56, 57, 59–60, 82
  in U.S. electrical power grid, 57
degrees of separation (diameter), 38,
      41–44, *43*, 53, 54–55, 211n
  in actors' networks, 56
  of corporate boards of directors, 117
  of English language, 88
  in food webs, 150, 151, 153
  of global ecosystem, 150–51
  in human brain, 65
  in Internet, 81
  in living cell, 135
  in random networks, 56, 57, 60,
      69, 131, 199
  of Silwood Park food web, 150
  of small-world graphs, 54, 55,
      58–59, 60
  see also "six degrees of separation"

dendrites, 64
"denial of service" Internet attacks,
      128
Deutch, John, 129
diameter, *see* degrees of separation
diffusion-limited aggregation (DLA),
      103–5, *104*
Discovery Channel, 29
diseases, 133–37
  drug-resistant bacteria and,
      134–35, 171
  new antibiotics for, 134
  sexually transmitted, *see* sexually
      transmitted diseases
diseases, spreading of, 22, 33, 137,
      159, 170–83
  in aristocratic networks, 180–82,
      183
  computer modeling of, 177–80,
      205
  cross-species infections in, 173–74
  element of chance in, 177
  global, 171–72
  noninfectious elements in, 176–77,
      178, 181
  in ordered networks, 175–77, *176*
  small-world graphs and, 176–78,
      176, 180
  social networks and, 175–80
  tipping point in, 161–64, 166–68,
      167, 175, 178, 180–82
  see also AIDS
distributed networks, 78–82, *79*
  fishnet, 78–80
  hiercharchical decentralized,
      80–81, 82
distribution of wealth, 18, 19,
      188–96, 207–8
  computer modeling of, 192–96
  condensed phase of, 195–96
  degrees of inequality in, 192–95
  80-20 principle in, 189
  government deregulation and, 194
  interpersonal transactions in,
      191–92, 193, 195

investment returns in, 191–93, 194, 195
Pareto's law of, 188–90, 192, 194, 195, 196, 198
political instability and, 196
power-law pattern in, 189, 193, 196
relative value of money in, 191
talent differential in, 190, 192
taxation in, 193–94, 196
tipping point of, 195
"trickle down" theory of, 194
washing-out effect in, 192–93
Don Juan de Marco, 14
Donohue, George L., 122
Dostoyevsky, Fyodor, 156
"dot com" Internet stocks, 160
Drucker, Peter, 76–77
drug-resistant bacteria, 134–35, 171

earthquakes, 100
eBay, 77, 128
e-commerce, 77–78
economics, 20, 47, 159–60, 187–96, 198, 199–201
   behavioral, 188
   free-market theories of, 187–88, 194
   social capital in, 201, 202–4, 205, 207
   wealth concentration in, *see* distribution of wealth
economy, 190–91
   business links of, 15
   threats to, 128, 129
ecosystems, 20, 22, 137, 138–55
   biomass of, 146
   connectors in, 149, 151–52, 153–54
   conservation of, 152
   diversity of, 146
   foreign species invasions of, 152
   global, 12, 143, 150–51, 184
   keystone species in, 153–54
   models of, 144–45
   population fluctuations in, 147–48
   as random networks, 144–45, 146
   simplified, 144, 152

species extinctions in, 143, 144, 145, 147, 148, 152, 153
   stability of, 143–49, 152, 154
ecosystems, food webs of, 15, 60, 138–55, 197
   as aristocratic networks, 149–52
   complexity of, 141–46, *142*
   degrees of separation in, 150, 151, 153
   growth of, 119
   marine, fishing industry and, 16, *17*, 138–43
   North Atlantic, 140–43, *142*, 218*n*
   power-law pattern of, 151
   predator-prey relationship in, 88, 143, 144, 145, 148, 152, 154
   strong ties in, 145, 148–49, 152
   weak ties in, 145–49, 151, 152, 153
Eddington, Rod, 122
Edling, Christofer, 113–14
egalitarian networks, 118–20, 123, 124–25, 126, 137
   neural networks as, 119, 125
Egypt, 21
Eigen, Manfred, 61
80-20 principle, 189
Einstein, Albert, 51, 72, 90, 108
Eisenhower, Dwight, 73–74
electrical power grid, U.S., 14–15, 22, 128, 129, 132
   degree of clustering in, 57
   as egalitarian network, 119 20, 125
   interconnections in, 56–57
Elements (Euclid), 71
Elton, Charles, 144
emergence, concept of, 198, 207
England, 21, 64–65, 178–79
   air traffic in, 122
   distribution of wealth in, 196
   foot-and-mouth disease epidemic in, 175
   rioting in, 106–7
   Silwood park research area of, 149–50, 151, 153–54

English language, 88, 119
epidemiology, 47, 161–64, 180,
    182–83
  see also diseases, spread of
Erdös, Paul, 34–35, 36–37, 39, 40, 145
erosion, 101–2, *102*
Escherichia coli, 87, 97, 133, 134, 135
  0155 and 0157:H7, 171
Euclid, 71–72
evolution, biological, 87, 90, 91–92,
    198
extinctions:
  mass, 18
  species, 143, 144, 145, 147, 148,
    152, 153

factory workers, 202–3
Faloutsos, Michalis, Petros, and
    Christos, 81–82, 83, 84, *85,*
    213*n*
Faraday, Michael, 94
Federal Aviation Administration, 122
Fella River network, 98, *98*
Ferrer i Cancho, Ramon, 88
Fiat, 206–7
fireflies, synchronous flashing by,
    48–49, 50, 57–59, 68–69, 208,
    211*n*
firewalls, 132
fishing industry, 138–43
  Canadian, 140–41, 143, 153, 218*n*
  commercial whaling and, 138–40,
    141
  Japanese, 138–40, 153
  South African, 16, *17,* 218*n*
fluid physics, 93–94, 214*n*
flu virus, 162, 175, 176
Food and Agriculture Organization
    (FAO), UN, 139
food poisoning, 134
food webs, *see* ecosystems, food webs
    of
foot-and-mouth disease, 175
fractals, 103, 104–5
France, 178

factory workers in, 202–3
free-market economic theories,
    187–88, 194
friendship networks, 65, 199, 204
  degree of clustering in, 38–40, 65
  graphs of, 31–32, *31, 32*
  weak ties in, 45–46
Fukayama, Francis, 201, 202
functional magnetic resonance, 63
Futurist, 31

Gabriel, Ralph, 89
Galbraith, John Kenneth, 189
Gall, Franz Joseph, 61–63, 212*n*
Gans, Herbert, 204–5
Gates, Bill, 190
genes, 133, 209*n*
  memes and, 160
genetic mutations, 135
  of bacteria, 171
  of viruses, 174–75
genomes:
  brewer's yeast, 135–37, *136*
  human, 16, 20
  plant, 16, 20
geometry, 71–72
  non-Euclidean, 51, 72
  see also patterns, geometrical
Germany, 13–14, 67–68, 90–91
  factory workers in, 202–3
Gerstner, Lou, 77
Gibson, Steve, 129
Giles, Lee, 109–10
Gladwell, Malcolm, 114, 158–59, 162,
    164, 168
global politics, 21
Gluckman, Max, 138
Google, 109, 110
Gould, Stephen Jay, 91
Grameen banking, 203
Granovetter, Mark, 40, 41–46, 47, 51,
    65, 118–19, 148, 199, 200,
    204, 205
  on group behavior, 107–8,
    214*n*–15*n*

graphs, 13, 35–40, 52, 56
  of friendship networks, 31–32, *31*,
    32
  mathematical, 30–31, *31*, *32*
  ordered vs. random, *see* ordered
    networks; random networks
  of road networks, 35–36
  of social bridges, 41–44
  of strong ties, 41–42, *42*
  see also power-law pattern; small-
    world graphs
Grattan-Guinness, Ivor, 23
Greenpeace, 139
Grossman, Jerrold, 210*n*
group behavior, 106–8, 159–60,
    214*n*–15*n*
groupthink, 115–16
"grunge" music, 159
Guare, John, 13

haddock, 139
hake, 16, *17*
hand clapping, synchronized, 49,
    211*n*
harp seals, 140–41, 218*n*
Harris, Leslie, 140
Hartsfield-Atlanta International
    Airport, 121
Harvard University, 25
Hastings, Alan, 147–48
Heathrow Airport, 122
hexagonal patterns, 92–94, *93*, 95, *95*,
    96, *97*
hippocampus, 63, 64, 66
historical sciences, 90, 103
historicism, 11–12
history, 89–92, 96–97, 119, 125, 190,
    198, 202
  contingency in, 90–92, 96, 100,
    102–3, 105
  methods of, 89
  river networks and, 98, 100–103
HIV virus, 162, 163, 171, 172–74, 181
  see also AIDS
Holland, 160

Hong Kong, 175
Hooper, Edward, 174
Horvath, W., 45–46
hubs, 86–87, 91, 113
  airports as, 114, 121–22, 123–24,
    125
  in aristocratic networks, 119, 123,
    124, 131–32
  computer-simulated attacks on, 132
  connectors as, 114–15, 119, 149
  of English language, 88
  of Internet, 82, 85, 86
  keystone species as, 153–54
  proteins as, 116–17, 134, 136–37
  superconnected, 123, 124, 149,
    151, 208
  of World Wide Web, 97, 123, 124
human genome, 16, 20
Human Genome Project, 16, 59
Hush Puppies shoes, 159, 164
Huxel, Gary, 147–48
Huygens, Christian, 49–50

"I LOVE YOU" computer virus, 128
Independence Day, 28
Industrial Revolution, 77
influences, spreading of, 32–33, 46,
    114, 117, 137, 158–61, 205
  see also diseases, spreading of; tip-
    ping point
information infrastructure, U.S.,
    127–33
  protection of, 132–33
  vulnerability of, 127–30, 134
  see also cyber-crime
Information Revolution, 76–77, 83
Institute of Medicine, U.S., 134
interconnections, 33, 56–57
International Whaling Commission,
    138, 139
Internet, 12, 21, 22, 60, 73–87, 91, 96,
    105, 108, 119, 133, 134, 135,
    197, 198, 208
  as aristocratic network, 125, 130,
    131

Internet *(continued)*
  complexity of, 28
  degree of clustering in, 82
  degree of separation in, 81
  "dot com" stocks and, 160
  e-commerce on, 77–78
  global topology of, 80–82, *81*
  growth of, 76, 78, 82, 110–11, 116,
    119
  as hierarchical decentralized dis-
    tributed network, 80–81, 82
  hubs of, 82, 85, 86
  impact of, 76–78
  interdomain level of, 213*n*
  modeling of, 126–27
  nodal links of, 82–86, *85*
  origin of, 73–76
  power-law pattern in, 83–85, *85*
  protocols of, 126–27
  redundancy of, 75
  routing tables of, 75
  scientific publishing on, 78
  structure of, 78–86, 87
  see also cyber-crime; World Wide
    Web
Internet Movie Database, 29, 56,
  210*n*
investment returns, 160
  in distribution of wealth, 191,
    192–93, 194
Islamism, 21
island habitats, small, 144
Italy, 98, *98*
  automakers in, 206–7

Janus, Irving, 115–16
Japan, 134
  fishing industry of, 138–40, 153
Jeong, Hawoong, 87, 130–33, *136*
job-search networking, 44–46, 204
Joffee, George, 21
Jouvenal, Bertrand de, 197
Jung, Carl, 40

Kagera Salient, 179

Kant, Immanuel, 19, 61
Kapitsa, Pyotr, 157
keystone species, 153–54
Kleinock, Leonard, 75
Klovdahl, Alden, 32
Koretz, Moissey, 156–57
Kuhn, Thomas, 47

Lago-Fernández, Luis, 69–70
Landau, Lev, 156–58, 164–66, 169,
  219*n*
languages, 60, 88, 91, 119, 197, 208
Latora, Vito, 65
Laughlin, Robert, 184
Laurent, Gilles, 68
Lawrence, Steve, 109–10
Lenin, Vladimir, 117–18
Leucippus, 89
life expectancy, 170
Liljeros, Fredrik, 113–14, 115
Little Rock Lake, food web of, 150,
  151, 153–54
Lobachevskii, Nickolai, 71–72
Locke, Richard, 206–7
locusts, brains of, 68, 69–70
lost-letter technique, 25

McCann, Kevin, 147–48, 152
magnetic resonance imaging, func-
  tional, 63
Malaysia, 59
Mann, Jonathan, 170, 172
Marchiori, Massimo, 65
Marx, Karl, 11–12
Mason, Sean, 136–37
mass extinctions, 18
mathematical games, 103–5
mathematical graphs, 30–33, *31, 32*
mathematicians' networks, 34–35
Max Planck Institute for Brain
  Research, 67–68
Mayor of Sunset Strip, 29
May, Robert, 144–45, 146–47
Medline, 116
memes, 160

Metacrawler, 110
methicillin, 134
Mexico, 189, 195
Mézard, Marc, 191–96
Michigan, University of, 117, 145–46
Milgram, Stanley, 25–28, 34
    letters experiment of, 13, 25–26,
        27–28, 29, 30, 41, 47, 51, 114
    lost-letter technique of, 25
    "obedience to authority" experi-
        ments of, 26–27, 200–201
mind, human, 61
minke whales, 138, 139
Minnesota, 145–46
Mirollo, Renato, 50, 57–58
Mississippi River, 97–98, 99, 100
Mizruchi, Mark, 118
Molotov, Vyacheslav, 157
monkeys, 172–74
    brain of, 64–65
Montoya, José, 150–51, 153–54
Motulsky, Arno, 174
My Dog Skip, 28, 29
Myers, Ransom, 140–41, 144–45

name recognition, 113, 116–17
Namhias, André, 174
Napoleon I, Emperor of France, 62
National Aeronautics and Space
        Administration (NASA), 74
National Institutes of Health, 116
National Security Agency, 129
Nature, 23–24, 47, 60, 78
NEC Research Institute, 109–10
nematode (Caenorhabditis elegans),
        14–15, 59–60, 119, 125
neural networks, 14–15, 59–60,
        64–66, 67, 197
    axons in, 64, 65
    dendrites in, 64
    as egalitarian networks, 119, 125
    timing in, 70
    see also brain; brain, human
neural synchrony, 49, 67–70
New Guinea, 48–49, 59

Newman, Mark, 87–88, 116
New York, N.Y., 159, 164
New York Times, 29–30
non-Euclidean geometry, 51, 72
North Atlantic Ocean, 140–43, 142,
        218n
Northwestern University, 134, 136
Norway, 95
Notre Dame University, 83–84, 111

"obedience to authority" experi-
        ments, 26–27, 200–201
O'Hare International Airport, 121,
        123
Oltavi, Zoltan N., 87
Oracle of Kevin Bacon, 28–29
ordered networks, 14, 40, 52–55, 53,
        64, 69, 70, 78–79, 82, 110
    degrees of clustering in, 51, 53–54
    degrees of separation in, 54, 55
    disease spreading in, 175–77, 176

Palmer, Arnold, 29
Papua, New Guinea, 48–49
Pareto, Vilfredo, 188–96, 198, 208
Pastor-Satorras, Romualdo, 181
patterns, geometrical, 92–97
    of Arctic rocks, 95, 96
    of heated water, 92–94, 93, 96, 97
    of shaken sand, 95, 95, 96
    of shaken water, 94–95, 96
permission marketing, 160–61
Perot, Ross, 159
pest invasions and outbreaks, 144
phase transitions, 157–58, 164–66,
        219n
Philippines, 21
phrenology, 62, 212n
physics, 89, 90, 154–55, 156–58, 207,
        208
    fluid, 93–94, 214n
    phase transitions in, 157–58,
        164–66, 219n
Pines, David, 184
pipeline control systems, 132

plant genomes, 16, 20
Plato, 71, 97
Platonic forms, 19
Poincaré, Henri, 11
polio vaccines, 174
Polya, George, 47
Popper, Karl, 11–12, 21
Postal Service, U.S., 21–22, 25, 128
Potterat, John, 161–62, 183
Poverty of Historicism, The (Popper),
    11–12
power-law pattern, 83–85, *85*, 86, 87,
    215*n*
  of diffusion-limited aggregation,
    103–5
  in distribution of wealth, 189, 193,
    196
  fat tails of, 84, 91, 115, 124, 189,
    193, 196
  of food webs, 151
  in preferential attachments, 113,
    114, 115
  of river networks, 99–100, 102–3
  of sexual contact links, 114
predator-prey relationship, 88, 143,
    144, 145, 148, 152, 154
preferential attachments, 106–20, 198
  in actors' networks, 111, 113
  in aristocratic vs. egalitarian net-
    works, 118–20
  connectors in, 114–15, 119, 120
  in corporate boards of directors,
    116–18
  diminishing returns of, 123–25
  in group behavior, 106–8
  in groupthink, 115–16
  mechanism of, 110–13
  in Milgram's letters experiment,
    114
  name recognition in, 113, 116–17
  to popular Web sites, 108–10, 112,
    115, 116, 119, 123, 208
  power-law pattern in, 113, 114,
    115
  in scientific research paper cita-

    tions, 111, 112, 119
  of scientific research paper collab-
    orators, 116, 119
  in sexual contact networks,
    113–15, 119
  in social networks, 113–18
presidential election of 1992, 159
Presley, Elvis, 28, 29
proteins, 16, 60, 133, *136*, 209*n*
  as hubs, 116–17, 134, 136–37
public health measures, 162, 164,
    170, 171, 180, 182–83
Public Library of Science, online,
    78
Pythagoras, 71, 97

quantum theory, 90

racial segregation, origin of, 185–86,
    186
RAND corporation, 74, 78, 129, 132
random networks, 14, 35–40, 51–55,
    57, 69–70
  computer-simulated attacks on,
    130–32
  degrees of clustering in, 38–40,
    51–52, 54, 56, 57, 59–60, 82
  degrees of separation in, 56, 57,
    60, 69, 131, 199
  ecosystems as, 144–45, 146
  percentage of links in, 37
  road networks as, 35–36
Rapoport, Anatol, 45–46
Rayleigh, Lord, 93–94, 95, 214*n*
Redner, Sidney, 87–88
reductionism, 15, 184–86, 207
redundancy, computer network, 130,
    131, 132
  of Internet, 75
Reeves, Keanu, 29
relativity, theory of, 51
Rényi, Alfréd, 35
Republic (Plato), 71
research papers, *see* scientific
    research papers

resource "engineering," 153
Rinaldo, Andrea, *98*, 101–3, *102*
rioting, 106–8, 214*n*–15*n*
river networks, 97–103, *98*, 113, 190
    catchment areas in, 99
    computer modeling of, 101–3, *102*
    erosion and, 101–2, *102*
    history and, *98*, 100–103
    power-law pattern of, 99–100,
        102–3
    self-similarity in, 103
River, The (Hooper), 174
road networks, 35–36, 119
rocks, Arctic, pattern made by, 95, *96*
Rodríguez-Iturbe, Ignacio, *98*, 101–3,
    102
Rumer, Yuri, 156–57
Russia, 189, 196

Saccharomyces cerevisiae, 135–37, *136*
sand, shaken, patterns made by, 95,
    95, *96*
Santa Fe Institute, 116
satellite program, U.S., 73–74
Saudi Arabia, 21
Saxenian, Analee, 206
Say One for Me, 29
"scale-free" networks, *112*, 115, 215*n*
Scannell, Jack, 64–65
Schelling, Thomas, 185–86, *186*
science, 51, 78, 89, 137, 198
    historical, 90, 103
    mathematical, 90
    reductionism in, 15, 184–86, 207
scientific publishing, 78
scientific research papers, 78
    citations in, 87, 111, 112, 119
    collaborators on, 87–88, 116, 119
scientific revolutions, 47
Scotch broom, 150
seals, 16, 140–41, 218*n*
search engines, 109–10
self-similarity, 103, 104–5
September 11 terrorist attacks,
    20–22, 122, 127, 128

sexual contact networks, 113–14,
    119, 180–82
sexually transmitted diseases, 180–83
    connectors and, 180–83
    core group in, 182–83
    syphilis, 161–62, 164, 176
    see also AIDS
Silicon Valley, 205–6
Silwood Park, food web of, 149–50,
    151, 153–54
Simon, Herbert, 12, 198
Singer, Wolf, 67–68
SIV viruses, 172–74
Six Degrees of Monica Lewinsky, 30
"six degrees of separation," 13–14,
    19–20, 25–26, 27–31, 38, 44,
    47, 51, 55, 118, 175
small pox virus, 170
small-world graphs, 14–15, 30,
    52–60, *53*, 82, 198
    computation and, 59–60
    degrees of separation in, 54, 55,
        58–59, 60
    disease spreading and, 176–78,
        176, 180
Smith, Adam, 187, 198
Smith, Will, 28
smurf attacks, 128
social bridges, 41–46, 47, 55, 118–19
social capital, 201, 202–4, 205, 207
social changes, 158–61, 164
social embeddedness, 201
social networks, 13–15, 22, 23–60, 72,
    91, 145, 148–49, 199–207
    acquaintances in, 30, 31–32, *31*, *32*,
        37, 38–40, 44, 114
    of actors, 28–30, 56
    authoritive relationships and,
        200–201
    blacks vs. whites in, 28
    coincidences in, 24, 25, 40, 197
    connectors in, 114–15, 119
    ethical values and, 199–200
    graphs of, *see* graphs
    of mathematicians, 34–35

social networks *(continued)*
  Milgram's letters experiment on,
      13, 25–26, 27–28, 29, 30, 41,
      47, 51, 114
  percentage of links in, 37
  preferential attachments in,
      113–18
  prevalence of trust in, 200, 201,
      202–4, 205, 206
  racial segregation and, 185–86, *186*
  social distances in, 41–44
  spreading through, 32-33, 46, 114,
      117, 137, 158-61, 205
      see also diseases, spreading
      of; tipping point
  of world populations, 23, 37–40,
      55, 118
Socrates, 71
Solé, Ricard, 88, 150–51, 153–54
Somalia, 21
sooty mangabeys, 173, 174
South Africa, 16, *17*, 218*n*
Soviet Union, 21, 73–74, 156–58
  wealth distribution after collapse
      of, 196
Speedway, 29
Spice Girls, 30
Sputnik satellite, 73–74
Stanford University, 76
Staphylococcus aureas, 134
Star Links, 29
Steenburgen, Mary, 30
Stewart, William H., 170
Stockholm University, 113–14
stock market, 188, 191
  price fluctuations on, 196
"Strength of Weak Ties, The" (Gra-
      novetter), 43–44
Strogatz, Steve, 13–15, 19, 23–24, 47,
      50, 52–60, 65, 68–69, 70, 78,
      82, 86, 91, 110, 111, 118–19,
      126, 180, 198, 211*n*
strong ties, 43, 44, 55, 199, 202
  in food webs, 145, 148–49, 152
  graphs of, 41–42, *42*

superfluids, 219*n*
Sweden, 113–14, 180
Symington, Stuart, 73
syphilis, 161–62, 164, 176

Tanzania, 171, 179
Tanzanian People's Defense Force
      (TPDF), 179
taxation, 193–94, 196
telephone network, U.S., 22, 74–75,
      76, 132
terrorist networks, 20–22, 127
Thailand, 59
"thing-in-itself" concept, 19
Thomson, D'Arcy Wentworth, 121
Tierney, John, 48
Tilman, David, 145–46
tipping points, 156–69, 185
  connectors and, 168–69
  in contact process games, 166–68,
      167
  critical state as, 166
  of disease spreading, 161–64,
      166–68, *167*, 175, 178, 180–82
  of distribution of wealth, 195
  in phase transitions, 157–58,
      164–66
  social changes and, 158–61, 164
Tipping Point, The (Gladwell), 114,
      158–59, 162, 164
Tjaden, Brett, 28–29
Tombor, Balint, 87
transportation networks, 119–20
  road networks, 35–36, 119
      see also air transportation net-
      works
travel, 121
Treponema pallidum, 162
"trickle down" theory, 194
tropical forests, 152
Trotsky, Leon, 100
trust, prevalence of, 200, 201, 202–4,
      205, 206
tulip mania, 160
Twain, Mark, 106

Uganda, 171, 179
United Nations, 171
    Economics and Social Affairs
        Department of, 24
    Food and Agriculture Organiza-
        tion of, 139
upheavals, tumultuous, 20–21
Up in Smoke, 29

vancomycin, 134
Vespignani, Alessandro, 181
Virginia, University of, 28
viscosity, 93, 94

Wagner, Robert, 29
Wain-Hobson, Simon, 178–79
water, patterns made by, 92–95, *93*,
        96, 97
Watts, Duncan, 13–15, 19, 23–24, 47,
        49, 50–60, 65, 68–69, 70, 78,
        82, 86, 91, 110, 111, 118–19,
        126, 180, 198, 211*n*
weak ties, 41–47, 51, 145, 199
    absence of, 204–7
    in food webs, 145–49, 151, 152,
        153
    in friendship networks, 45–46
    in job-search networking, 44–46,
        204
    as social bridges, 42–46, 47, 55
    see also strong ties
wealth, distribution of, *see* distribu-
        tion of wealth
Wealth of Nations, The (Smith), 187
Wendt, George, 30
Wesson, Glenn, 28–29

whaling, commercial, 138–40, 141
Whitehead, Alfred North, 90, 103
Whyman, Mark, 106
Williams, Richard, 150, 151
Winkler, Peter, 34
Wisconsin, 150
working-class culture, 204–5
World Health Organization, 171
World Wide Web, 15, 19, 76, 78,
        108–11, 133, 197
    "diameter" of, 85–86
    growth of, 83, 86, 109–11, 119
    hubs of, 97, 123, 124
    hypertext links of, 83, 108–9
    new Web pages on, 108–10
    Oracle of Kevin Bacon, 28–29
    preferential attachment to popular
        sites on, 108–10, 112, 115,
        116, 119, 123, 208
    search engines for, 109–10
    smurf attacks on, 128
    structure of, 83–86, 87

Xerox Internet Ecologies Division,
        76, 213*n*

Yahoo, 109, 123, 128
yeasts, 135–37, *136*
Yemen, 21
Yodzis, Peter, 16, 143, 145, 218*n*
Ythan River estuary, food web of,
        150, 151, 153–54
Yunas, Muhammad, 203

Zanette, Damian, 176–80, 205
Zeit, Die, 13–14